TUT

19
Maths for Chemists

Volume 2

Power Series, Complex Numbers and Linear Algebra

MARTIN COCKETT & GRAHAM DOGGETT
University of York

RS•C
ROYAL SOCIETY OF CHEMISTRY

ISBN 0-85404-495-7

A catalogue record for this book is available from the British Library

Published by The Royal Society of Chemistry, Thomas Graham House, Science Park, Milton Road, Cambridge CB4 0WF, UK
Registered Charity No. 207890
For further information see our web site at www.rsc.org

Typeset in Great Britain by Alden Bookset, Northampton
Printed and bound in Italy by Rotolito Lombarda

Preface

These two introductory texts provide a sound foundation in the key mathematical topics required for degree level chemistry courses. While they are primarily aimed at students with limited backgrounds in mathematics, the texts should prove accessible and useful to all chemistry undergraduates. We have chosen from the outset to place the mathematics in a chemical context – a challenging approach because the context can often make the problem appear more difficult than it actually is! However, it is equally important to convince students of the relevance of mathematics in all branches of chemistry. Our approach links mathematical principles with the chemical context, by introducing the basic concepts first, and then demonstrates how they translate into a chemical setting.

Historically, physical chemistry has been seen as the target for mathematical support; however, in all branches of chemistry – be they the more traditional areas of inorganic, organic and physical, or the newer areas of biochemistry, analytical and environmental chemistry – mathematical tools are required to build models of varying degrees of complexity, in order to develop a language for providing insight and understanding together with, ideally, some predictive capability.

Since the target student readership possesses a wide range of mathematical experience, we have created a course of study in which selected key topics are treated without going too far into the finer mathematical details. The first two chapters of Volume 1 focus on numbers, algebra and functions in some detail, as these topics form an important foundation for further mathematical developments in calculus, and for working with quantitative models in chemistry. There then follow chapters on limits, differential calculus, differentials and integral calculus. Volume 2 covers power series, complex numbers, and the properties and applications of determinants, matrices and vectors. We avoid discussing the statistical treatment of error analysis, in part because of the limitations imposed by the format of this series of tutorial texts, but also because the procedures used in the processing of experimental results are commonly provided by departments of chemistry as part of their programme of practical chemistry courses. However, the propagation of errors, resulting from the use of formulae, forms part of the chapter on differentials in Volume 1.

Martin Cockett
Graham Doggett
York

TUTORIAL CHEMISTRY TEXTS

EDITOR-IN-CHIEF

Professor E W Abel

EXECUTIVE EDITORS

Professor A G Davies
Professor D Phillips
Professor J D Woollins

EDUCATIONAL CONSULTANT

Mr M Berry

This series of books consists of short, single-topic or modular texts, concentrating on the fundamental areas of chemistry taught in undergraduate science courses. Each book provides a concise account of the basic principles underlying a given subject, embodying an independent-learning philosophy and including worked examples. The one topic, one book approach ensures that the series is adaptable to chemistry courses across a variety of institutions.

TITLES IN THE SERIES

Stereochemistry *D G Morris*
Reactions and Characterization of Solids *S E Dann*
Main Group Chemistry *W Henderson*
d- and f-Block Chemistry *C J Jones*
Structure and Bonding *J Barrett*
Functional Group Chemistry *J R Hanson*
Organotransition Metal Chemistry *A F Hill*
Heterocyclic Chemistry *M Sainsbury*
Atomic Structure and Periodicity *J Barrett*
Thermodynamics and Statistical Mechanics *J M Seddon and J D Gale*
Basic Atomic and Molecular Spectroscopy *J M Hollas*
Organic Synthetic Methods *J R Hanson*
Aromatic Chemistry *J D Hepworth, D R Waring and M J Waring*
Quantum Mechanics for Chemists *D O Hayward*
Peptides and Proteins *S Doonan*
Reaction Kinetics *M Robson Wright*
Natural Products: The Secondary Metabolites *J R Hanson*
Maths for Chemists, Volume 1, Numbers, Functions and Calculus *M Cockett and G Doggett*
Maths for Chemists, Volume 2, Power Series, Complex Numbers and Linear Algebra *M Cockett and G Doggett*

FORTHCOMING TITLES

Mechanisms in Organic Reactions
Molecular Interactions
Lanthanide and Actinde Elements
Maths for Chemists
Bioinorganic Chemistry
Chemistry of Solid Surfaces
Biology for Chemists
Multi-element NMR
EPR Spectroscopy
Biophysical Chemistry

Further information about this series is available at www.rsc.org/tct

Order and enquiries should be sent to:
Sales and Customer Care, Royal Society of Chemistry, Thomas Graham House, Science Park, Milton Road, Cambridge CB4 0WF, UK

Tel: +44 1223 432360; Fax: +44 1223 426017; Email: sales@rsc.org

Contents

Symbols

$>$	greater than	$\equiv$	equivalent to
$\geqslant$	greater than or equal to	$\propto$	proportionality
$\gg$	much greater than	$=$	equality
$<$	less than	∞	infinity
$\leqslant$	less than or equal to	$\sum$	summation sign
$\ll$	much less than	$\prod$	product sign
$/$ or $\div$	division	!	factorial
$\neq$	not equal to	{ }	braces
$\cong$ or $\approx$	approximately equal to	[ ]	brackets
$\Rightarrow$	implies	( )	parentheses

1
Power Series

In Chapter 2 of Volume 1 we saw that a **polynomial function**, $P(x)$, of degree n in the independent variable, x, is a finite sum of the form:

$$P(x) = c_0 + c_1x + c_2x^2 + \cdots + c_nx^n \tag{1.1}$$

Such polynomial functions have as **domain** the set of all real (finite) numbers; in other words, they yield a finite result for any real value of the independent variable x. For example, the polynomial function:

$$P(x) = 1 + x + x^2 + x^3 \tag{1.2}$$

will have a finite value for any real number x because each term in the polynomial will also have a finite value.

An **infinite series** is very similar in form to a polynomial except that it does not terminate at a particular power of x and, as a result, is an example of a **power series**:

$$p(x) = c_0 + c_1x + c_2x^2 + \cdots + c_nx^n + \cdots \tag{1.3}$$

One important consequence of this lack of termination is that we need to specify a domain which includes only those real numbers, x, for which $p(x)$ is finite. For example, the power series:

$$p(x) = 1 + x + x^2 + x^3 + \cdots \tag{1.4}$$

does not yield a finite result for $x > 1$, or for $x < -1$, because in the former case the summation of terms increases without limit, and in the latter it oscillates between increasingly large positive and negative numbers as more and more terms are included. Try this for yourself by substituting the numbers $x = 2$ and $x = -2$ into equation (1.4) and then $x = 0.5$ and $x = -0.5$ and observe what happens to the sum as more and more terms are included.

Power series are useful in chemistry (as well as in physics, engineering and mathematics) for a number of reasons:

- Firstly, they provide a means to formulate alternative representations of transcendental functions such as the exponential, logarithm and trigonometric functions introduced in Chapter 2 of Volume 1.
- Secondly, as a direct result of the above, they also allow us to investigate how an equation describing some physical property behaves for small (or large) values for one of the independent variables.

For example, the radial part of the 3s atomic orbital for hydrogen has the same form as the expression:

$$R(x) = N(2x^2 - 18x + 27)e^{-x/3} \tag{1.5}$$

If we replace the exponential part of the function, $e^{-x/3}$, with the first two terms of its power series expansion $(1-x/3)$, we obtain a polynomial approximation to the radial function given by:

$$R(x) = N(2x^2 - 18x + 27)(1 - x/3) \approx N(8x^2 - 2x^3/3 - 27x + 27) \tag{1.6}$$

We can see how well equation (1.6) approximates equation (1.5) by comparing plots of the two functions in the range $0 \leqslant x \leqslant 20$, shown in Figure 1.1.

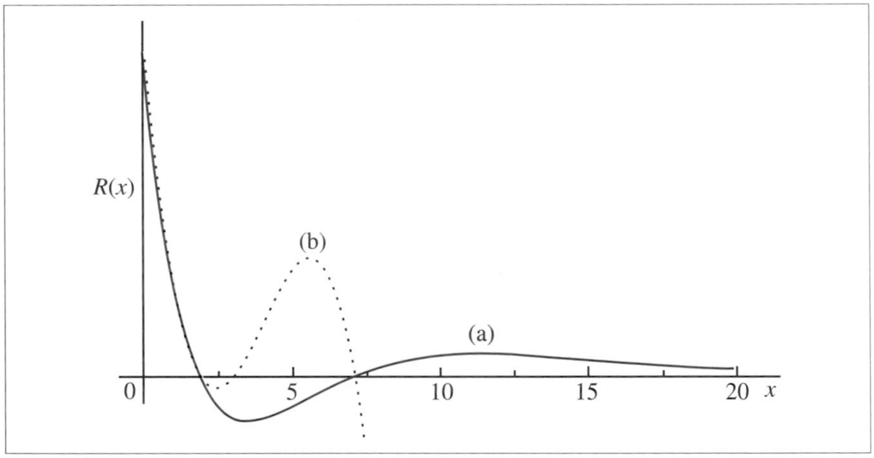

Figure 1.1 A comparison of (a) the 3s radial function, $R(x)$, of the hydrogen atom with (b) an approximation obtained by substituting a two-term expansion of the exponential part of the function

In this example, the polynomial approximation to the form of the radial wave function gives an excellent fit for small values of x (*i.e.* close to the nucleus), but it fails to reproduce even one **radial node** [a value of x for which $R(x) = 0$].

- Thirdly, power series are used when we do not know the formula of association between one property and another. It is usual in such situations to use a power series to describe the formula of association, and to fit the series to known experimental data by varying the appropriate coefficients in an iterative manner. For example, the parameters α, β, γ in the polynomial expression:

$$C_p = g(T), \text{ where } g(T) = \alpha + \beta T + \gamma T^2$$

describing the temperature dependence of the specific heat capacity of a substance at constant pressure, C_p, may be found by fitting measured values of C_p over a range of temperatures to this equation.

Much of this chapter is concerned with a discussion of power series, but before we go into detail we consider the general concepts of sequences and series, both finite and infinite.

Aims

By the end of this chapter, you should be comfortable with the idea that functions can be represented in series form and be able to:

- Understand the distinction between Maclaurin and Taylor series expansions and appreciate when one or the other is the more appropriate
- Understand what factors influence the accuracy of a given power series expansion
- Determine the values of x for which the power series is useful (the interval of convergence)
- Understand why the interval of convergence may differ from the domain of the original function
- Manipulate power series to obtain series for new functions
- Apply some of the ideas explored in this chapter to probe the limiting behaviour of functions for increasingly large or small values of the independent variable

1.1 Sequences

A **sequence** is simply a list of terms:

$$u_1,\ u_2,\ u_3, \ldots \tag{1.7}$$

each of which is defined by a formula or **prescription**. The sequence may be finite or infinite, depending on whether it terminates at u_n, or continues indefinitely. Furthermore, the sequence u_1, u_2, u_3, ... represents a function, with a domain specified either as a **subset** of the positive integers or all positive integers.

1.1.1 Finite Sequences

The Geometric Progression

The numbers:

$$1, 2, 4, 8, \ldots, 256$$

form a finite sequence generated by the **general term**:

$$u_r = 2^r, \text{ where } r = 0, 1, 2, 3, \ldots, 8 \tag{1.8}$$

Here, the formula is just 2 raised to a power, the value of which is defined by each element of the domain. Notice that the use of r as a **counting index** is arbitrary: any other appropriate letter (with the exception of u which we have used already) would do. A counting index such as r is often termed a **dummy index**. An alternative way of generating this sequence is accomplished using a **recurrence relation** as the prescription, where each successive term is obtained from the previous term. For example, the sequence given in equation (1.8) can alternatively be expressed as:

$$u_r = \begin{cases} 1, & r = 0 \\ 2u_{r-1}, & r = 1, 2, 3, \ldots, 8 \end{cases} \tag{1.9}$$

which simply means that, starting from 1 as the first term, each successive term is obtained by multiplying the previous term by 2.

The finite sequence in equation (1.8) is an example of a **geometric progression**, having the general form:

$$a,\ ax,\ ax^2,\ ax^3,\ \ldots,\ ax^8 \tag{1.10}$$

In the case of the geometric progression defined by equation (1.8), $a = 1$, $x = 2$ and $u_r = ax^r$ for $r = 0, 1, 2, \ldots$

Arithmetic Progression

Consider the sequence of odd positive numbers 1, 3, 5, 7, ... , 31, which can be expressed either in terms of the general term:

$$u_r = 2r - 1,\ r = 1, 2, 3, \ldots, 16 \tag{1.11}$$

or as a recurrence relation expressed in prescription form:

$$u_r = \begin{cases} 1, & r = 1 \\ u_{r-1} + 2, & r = 2, 3, \ldots, 8 \end{cases} \tag{1.12}$$

This finite sequence is an example of an **arithmetic progression**, because each successive term is given by a sum having the general form:

$$a,\ a + d,\ a + 2d,\ a + 3d,\ \ldots,\ a + nd \tag{1.13}$$

where, in this example, $a = 1$, $d = 2$.

1.1.2 Sequences of Indefinite Length

In the sequence given in equation (1.13), the magnitudes of successive terms progressively increase. Some sequences, however, have the property that as the number of terms increases, the values of successive terms appear to be approaching a limiting value. For example, the terms in the **harmonic sequence**:

$$1, \frac{1}{2}, \frac{1}{3}, \frac{1}{4}, \ldots, \frac{1}{n}, \ldots, \tag{1.14}$$

where $u_r = 1/r, r = 1, 2, 3, \ldots,$ decrease in magnitude as r increases, and approach zero as r tends to infinity. Thus, we can define the **limit** of the sequence as:

$$\lim_{r \to \infty} (u_r) = \lim_{r \to \infty} \frac{1}{r} = 0 \tag{1.15}$$

If the limit of a sequence is a single finite value, say m, then:

$$\lim_{r \to \infty} (u_r) = m \tag{1.16}$$

and the sequence is said to **converge** to the limit m; however, if this is not the case, then the sequence is said to **diverge**. Thus, for the arithmetic progression defined in equation (1.13), the magnitudes of successive values in the sequence increase without limit and the sequence diverges. In contrast, the geometric progression in equation (1.10) will converge to a limiting value of zero if $-1 < x < 1$.

Problem 1.1

Test each of the following sequences for convergence. Where convergence occurs, give the limiting value:

(a) $u_r = \frac{1}{2^r}$, $r = 0, 1, 2, \ldots$

(b) $u_n = \frac{n-1}{2n}$, $n = 1, 2, \ldots$

(c) $u_r = \cos r\pi$, $r = 0, 1, 2, \ldots$

1.1.3 Functions Revisited

In our discussion of algebraic manipulation in Chapter 1 of Volume 1, we used the three-spin model for counting the various permitted orientations of three spin-$\frac{1}{2}$ nuclei. If we focus on the number of arrangements where r nuclei are in the spin-up state, then we see that there is only one arrangement where none of the nuclei has spin up; three where one nucleus has spin up; three where two nuclei have spin-up; and one where all three nuclei have spin up. Thus, we can define the sequence 1, 3, 3, 1, where the general term is given by $u_r = \frac{3!}{(3-r)!r!}$, where $r = 0, 1, 2, 3$. In general, the number of ways of selecting r specified objects from n objects is given by the expression ${}^nC_r = \frac{n!}{(n-r)!r!}$. In this example there are three nuclei and so $n = 3$, and there are ${}^3C_r$ (where $r = 0$, 1, 2, 3) ways in which 0, 1, 2 and 3 nuclei have spin up.

Problem 1.2

Find the sequence that represents the number of arrangements of six spin-$\frac{1}{2}$ nuclei, with r spin-up arrangements, where r now runs from 0 to 6.

1.2 Finite Series

For any sequence of terms u_1, u_2, u_3, ... , we can form a **finite series** by summing the terms in the sequence up to and including the nth term:

We first met the summation notation $\sum$ in Chapter 1 of Volume 1. In this example, the counting index, r, takes values from 1 to n.

$$S_n = u_1 + u_2 + u_3 + \cdots + u_n = \sum_{r=1}^{n} u_r \tag{1.17}$$

For example, the sum of the first n terms in the series obtained from the sequence defined by equation (1.8) is given by:

$$S_n = 1 + 2 + 2^2 + \cdots + 2^{n-1} \tag{1.18}$$

Evaluating this sum for $n = 1, 2, 3, 4, 5$ yields the sequence of **partial sums**:

$$S_1 = 1,\ S_2 = 3,\ S_3 = 7,\ S_4 = 15 \text{ and } S_5 = 31 \tag{1.19}$$

If we now look closely at this new sequence of partial sums, we may be able to deduce that the sum of the first n terms is $S_n = 2^n - 1$. In general, for a **geometric series** obtained by summing the members of the geometric progression, defined by equation (1.10), the sum of the first n terms is given by:

$$S_n = a + ax + ax^2 + \cdots ax^{n-1} \\ = a\left(\frac{1-x^n}{1-x}\right) \tag{1.20}$$

Problem 1.3

For the geometric series obtained by summing the first n terms of the geometric progression in equation (1.8), use equation (1.20) and appropriate values of a and x given in equation (1.10) to confirm that the sum of the first n terms is 2^n-1.

1.3 Infinite Series

We can also form an **infinite series** from a sequence by extending the range of the dummy index to an infinite number of terms:

$$S = u_1 + u_2 + u_3 + \cdots = \sum_{r=1}^{\infty} u_r \tag{1.21}$$

The summation of a finite series will always yield a finite result, but the summation of an infinite series needs careful examination to confirm that the addition of successive terms leads to a finite result, *i.e.* the series converges. It is important not to confuse the notion of **convergence** as applied to a series with that applied to a sequence. For example, the harmonic *sequence* given by equation (1.14) converges to the limit zero. However, somewhat surprisingly, the **harmonic series**:

$$S = \sum_{r=1}^{\infty} \frac{1}{r} = 1 + \frac{1}{2} + \frac{1}{3} + \frac{1}{4} + \cdots \tag{1.22}$$

does not yield a finite sum, S, and consequently does not converge. In other words, the sum of the series increases without limit as the number of terms in the series increases, even though the values of successive terms converge to zero. We can see more easily how this is true by breaking down the series into a sum of partial sums:

$$S = 1 + \frac{1}{2} + \left(\frac{1}{3} + \frac{1}{4}\right) + \left(\frac{1}{5} + \frac{1}{6} + \frac{1}{7} + \frac{1}{8}\right) + \cdots \tag{1.23}$$

Here, each successive sum of terms in parentheses will always be greater than $\frac{1}{2}$. For example:

$$\left(\frac{1}{3}+\frac{1}{4}\right) > \left(\frac{1}{4}+\frac{1}{4}\right) \quad \text{and} \quad \left(\frac{1}{5}+\frac{1}{6}+\frac{1}{7}+\frac{1}{8}\right) > \left(\frac{1}{8}+\frac{1}{8}+\frac{1}{8}+\frac{1}{8}\right) \qquad (1.24)$$

and, because this is an infinite series, it follows that the sum increases without limit:

$$S > 1+\frac{1}{2}+\frac{1}{2}+\frac{1}{2}+\cdots \qquad (1.25)$$

1.3.1 π Revisited: the Rate of Convergence of an Infinite Series

In Chapter 1 of Volume 1 we saw that the irrational number, π, can be calculated from the sum of an infinite series. One example given involved the sum of the inverses of the squares of all positive integers:

$$\frac{\pi^2}{6} = \sum_{r=1}^{\infty}\frac{1}{r^2} = 1+\frac{1}{2^2}+\frac{1}{3^2}+\frac{1}{4^2}+\frac{1}{5^2}+\cdots+\frac{1}{n^2}+\cdots \qquad (1.26)$$

This series converges extremely slowly, requiring well over 600 terms to provide precision to the second decimal place; in order to achieve 100 decimal places for π, we would need more than 10^{50} terms! However, the alternative series:

$$\frac{\pi}{2} = \frac{1}{1}+\frac{1\times1}{1\times3}+\frac{1\times1\times2}{1\times3\times5}+\frac{1\times1\times2\times3}{1\times3\times5\times7}+\ldots \qquad (1.27)$$

converges more rapidly, achieving a precision to the second decimal place in a relatively brisk 10 terms.

1.3.2 Testing a Series for Convergence

The non-convergence of the harmonic series, discussed above, highlights the importance of testing whether a particular series is convergent or divergent. For a series given by:

$$\sum_{r=1}^{\infty} u_r = u_1 + u_2 + u_3 + \cdots \qquad (1.28)$$

the first, and necessary, condition needed to ensure convergence is that $\lim_{r\to\infty} u_r = 0$. If this condition is satisfied (as it is in the series above for determining π), we can then proceed to test the series further for convergence. It should be emphasized, however, that satisfying this first condition does not necessarily imply that the series converges (*i.e.* we say that the first condition is not sufficient). For example, although the harmonic sequence is an example of one for which u_r tends to zero as $r \to \infty$, the corresponding harmonic series is not convergent.

The Ratio Test

A number of tests are available for confirming the convergence, or otherwise, of a given series. The test for **absolute convergence** is the simplest, and is carried out using the **ratio test**.

For successive terms in a series, u_r and u_{r+1}, the series:

- converges if $\lim_{r \to \infty} \left| \frac{u_{r+1}}{u_r} \right| < 1$ (1.29)
- diverges if $\lim_{r \to \infty} \left| \frac{u_{r+1}}{u_r} \right| > 1$ (1.30)

If, however, $\lim_{r \to \infty} \left| \frac{u_{r+1}}{u_r} \right| = 1$, then the series may either converge or diverge, and further tests are necessary.

The Infinite Geometric Series

The form of the geometric series in equation (1.20) generalizes to the form of equation (1.21) where, now;

$$S_n = a + ax + ax^2 + \cdots$$
$$= \sum_{r=1}^{\infty} ax^{r-1}$$

and $u_r = ax^{r-1}$. On using the ratio test in equation (1.29), we find that the series converges if

$$\lim_{r \to \infty} \left| \frac{ax^r}{ax^{r-1}} \right| < 1$$

That is, when $|x < 1|$. This constraint on the permitted values of x, for which the series converges, defines the **interval of convergence**.

Worked Problem 1.1

Q (a) Give the forms of u_{r+1} and u_r for the geometric series $1 + x + x^2 + x^3 + \cdots$

(b) Use the ratio test to establish that the series converges and find the interval of convergence.

(c) Given $x = 0.27$, calculate the sum of the series to two decimal places.

A (a) Since the first term in the series is a constant, we define the rth term as $u_r = x^{r-1}$ and the $(r+1)$th term as $u_{r+1} = x^r$.
(b) The ratio test yields:

$$\lim_{r\to\infty}\left|\frac{u_{r+1}}{u_r}\right| = \lim_{r\to\infty}\left|\frac{x^r}{x^{r-1}}\right| = |x|$$

and so the series converges if $|x| < 1$ and diverges if $|x| > 1$. If $x = \pm 1$, then, as we saw earlier, further tests are required to establish whether the series converges or diverges at these end points. However, in this case, we can see by inspection that for $x = +1$ the sum of the first r terms will be r and thus increases without limit as $r \to \infty$. For $x = -1$, the sum oscillates between 0 and 1, depending on whether r is even or odd. In both cases, a finite sum is not obtained as $r \to \infty$, and we can say that the series fails to converge for $x = \pm 1$. Consequently, the series converges if x takes the values $-1 < x < 1$: an inequality that defines the interval of convergence.
(c) Let S_n designate the sum of the first n terms of the geometric series. Table 1.1 summarizes the values of S_n, and the incremental changes for $n = 1, 2, \ldots, 8$, using $x = 0.27$. We can see from the table that in order to specify the sum to a given number of decimal places, we have to compute its value to one more decimal place than required, in case rounding up is necessary. In this case, convergence to two decimal places yields a sum of 1.370 at $n = 6$.

Table 1.1 Numerical summation of the geometric series $\sum_{r=1}^{n} x^{r-1}$

n	1	2	3	4	5	6	7	8
S_n	1.0	1.27	1.3429	1.3626	1.3679	1.3693	1.3697	1.3698
ΔS_n	–	0.27	0.0729	0.0197	0.0053	0.0014	0.0004	0.0001

Problem 1.4

For each of the following infinite series, use the ratio test to establish the interval of convergence:

(a) $S = 1 + 2x + 3x^2 + 4x^3 + \cdots$

(b) $S = 1 - x + \frac{x^2}{2!} - \frac{x^3}{3!} + \frac{x^4}{4!} - \cdots + (-1)^{r-1}\frac{x^{r-1}}{(r-1)!} + \cdots$

(c) $S = 1 + \frac{x^2}{2} - \frac{x^4}{4} + \cdots + (-1)^{r-1}\frac{x^{2r-2}}{2r-2}$

Hint: for part (a), you will need to find the general term first.

In general, it may not be possible to specify the value of the sum, S, in terms of x. Instead, we chose a value of x, and compute the sum to a given number of significant figures or decimal places.

1.4 Power Series as Representations of Functions

We have seen above that, for a geometric progression of the type given in equation (1.10), the sum of the first n terms is given by equation (1.20). Furthermore, for $a = 1$, we can see that:

$$\frac{1 - x^n}{1 - x} = 1 + x + x^2 + \cdots + x^{n-1} \tag{1.31}$$

This is an important expression because it allows us to see how a function such as $\frac{1 - x^n}{1 - x}$ can be represented by a polynomial of degree $n - 1$. However, if we now extend the progressionindefinitely to form the infinite geometric series $1 + x + x^2 + \cdots + x^{n-1} + \cdots$, we obtain an expansion of a function $\lim_{n\to\infty}\frac{1 - x^n}{1 - x}$ which converges only for values of x in the range $-1 < x < 1$ (see Worked Problem 1.1). If we now evaluate the limit as $n \to \infty$, for any x in the interval of convergence $-1 < x < 1$, we obtain:

$$\lim_{n\to\infty}\frac{1 - x^n}{1 - x} = \frac{1 - 0}{1 - x} = \frac{1}{1 - x} \tag{1.32}$$

Power series are so-called because they are sums of powers of x with specified coefficients.

Note that in the limit $n \to \infty$, the term $x^n \to 0$ because we are restricting the values of x to the interval of convergence $-1 < x < 1$. We now see that the function $f(x) = \frac{1}{1-x}$ can be represented by the infinite series expansion $1 + x + x^2 + \cdots + x^{n-1} + \cdots$, which converges for $-1 < x < 1$. For all other values of x the expansion diverges.

The infinite geometric series is an example of a **power series** because it contains a sum of terms involving a systematic pattern of change in the power of x. In general, the simplest form of a power series is given by:

$$f(x) = c_0 + c_1 x + c_2 x^2 + c_3 x^3 + \cdots + c_n x^n + \cdots \tag{1.33}$$

where c_0, c_1, c_2, ... are coefficients and successive terms involve an increasing power of the independent variable, x. Such series involving simple powers of x are termed **Maclaurin series**. The more general **Taylor series** are similar in form, but involve powers of $(x - a)$:

$$f(x) = c_0 + c_1(x - a) + c_2(x - a)^2 + c_3(x - a)^3 + \cdots + c_n(x - a)^n + \cdots \tag{1.34}$$

where a is any number other than 0 (in which case, we revert to a Maclaurin series). The significance of the value of a is that it represents the point about which the function is expanded. Thus the Taylor series are expanded about the point $x = a$, while the Maclaurin series are simply expanded about the point $x = 0$. Maclaurin and Taylor series are used most frequently to provide alternative ways of representing many types of function. In addition, such series in truncated polynomial form provide an excellent tool for fitting experimental data, when there is no model formula available.

There are two important features associated with the generation of power series representations of functions. First, a value of x lying in the domain of the function must be chosen for the expansion point, a; second, the function must be infinitely differentiable at the chosen point in its domain. In other words, differentiation of the function must never yield a constant function because subsequent derivatives will be zero, and the series will be truncated to a polynomial of finite degree. The question as to whether the power series representation of a function has the same domain as the function itself is discussed in a later subsection. The next subsection is concerned with determining the coefficients, c_i, for the two kinds of power series used to represent some of the functions introduced in Chapter 2 of Volume 1.

1.4.1 The Maclaurin Series

Expansion About the Point $x = 0$

Let us start by using equation (1.33) as a model expression to generate a power series expansion for a function $f(x)$, assuming that the requirements given in the paragraph above are satisfied. In order to obtain the explicit form of the series, we need to find values for the coefficients c_0, c_1, c_2, ... This is achieved in the following series of steps.

The original function, and its first, second and third **derivatives** are:

$$f(x) = c_0 + c_1x + c_2x^2 + c_3x^3 + c_4x^4 + \cdots + c_nx^n + \cdots \tag{1.35}$$

$$f^{(1)}(x) = c_1 + 2c_2x + 3c_3x^2 + 4c_4x^3 + \cdots + nc_nx^{n-1} + \cdots \tag{1.36}$$

$$f^{(2)}(x) = 2c_2 + 2 \times 3c_3x + 3 \times 4c_4x^2 + \cdots + n(n-1)c_nx^{n-2} + \cdots \tag{1.37}$$

$$f^{(3)}(x) = 2 \times 3c_3 + 2 \times 3 \times 4c_4x + \cdots + n(n-1)(n-2)c_nx^{n-3} + \cdots \tag{1.38}$$

If we now substitute the expansion point, $x = 0$, into each of the above equations we obtain:

$$f(0) = c_0 \tag{1.39}$$

$$f^{(1)}(0) = c_1 \tag{1.40}$$

$$f^{(2)}(0) = 2c_2 = 2!c_2 \tag{1.41}$$

$$f^{(3)}(0) = 2 \times 3c_3 = 3!c_3 \tag{1.42}$$

and, by inspection, the nth derivative has the form:

$$f^{(n)}(0) = n!c_n \tag{1.43}$$

For clarity, we use a superscript containing a counting number in parentheses to denote a particular order of derivative.

If we now substitute the coefficients obtained from each of the expressions (1.39)–(1.43) into (1.35), we obtain the Maclaurin series for $f(x)$:

$$f(x) = f(0) + f^{(1)}(0)x + \frac{f^{(2)}0)}{2!}x^2 + \frac{f^{(3)}(0)}{3!}x^3 + \cdots + \frac{f^{(n)}(0)}{n!}x^n + \cdots \tag{1.44}$$

This series, which is generated by evaluating the function and its derivatives at the point $x = 0$, is valid only when the function and its derivatives exist at the point $x = 0$ and, furthermore, only if the function is infinitely differentiable.

The Maclaurin Series Expansion for e^x

The exponential function $f(x) = e^x$ is unique insofar as the function and all its derivatives are the same. Thus, since $f^{(n)}(x) = e^x$, for all n, we have:

$$f(0) = f^{(1)}(0) = f^{(2)}(0) = f^{(3)}(0) = \ldots f^{(n)}(0) = e^0 = 1 \tag{1.45}$$

and, using equation (1.44), we obtain:

$$f(x) = e^x = 1 + x + \frac{x^2}{2!} + \frac{x^3}{3!} + \cdots + \frac{x^{n-1}}{(n-1)!} + \cdots \tag{1.46}$$

in which the nth term is given explicitly.

Truncating the Exponential Power Series

For any power series expansion, the accuracy of a polynomial truncation depends upon the number of terms included in the expansion. Since it is impractical to include an infinite number of terms (at which point the precision is perfect), a compromise has to be made in choosing a sufficient number of terms to achieve the desired accuracy. However, in truncating a Maclaurin series, the chosen degree of polynomial is always going to best represent the function close to $x = 0$. The further away from $x = 0$, the worse the approximation becomes, and more terms are needed to compensate, a feature which is demonstrated nicely in Figure 1.2 and Table 1.2.

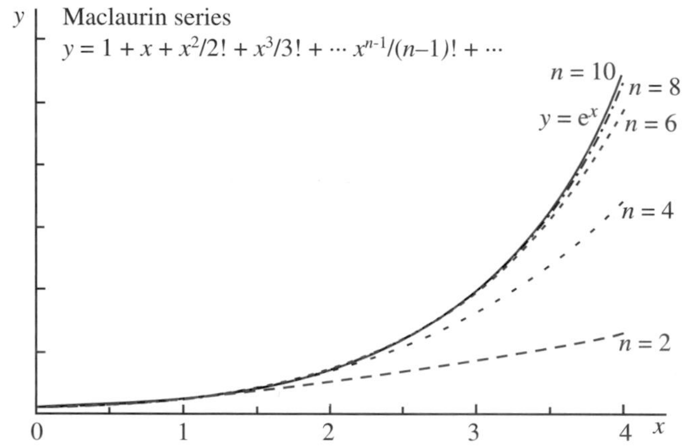

Figure 1.2 A comparison of the accuracy of polynomial approximations to the function $y = e^x$ ($x \geq 0$), using polynomials of degrees 2–10

Table 1.2 The accuracy of first, second and third degree polynomial approximations to the function $f(x) = e^x$

x	$1 + x$	$1 + x + \frac{x^2}{2!}$	$1 + x + \frac{x^2}{2!} + \frac{x^3}{3!}$	e^x
0	1	1	1	1
0.0001	1.0001	1.000100005	1.000100005	1.000100005
0.001	1.001	1.0010005	1.0010005	1.0010005
0.01	1.01	1.01005	1.01005017	1.01005017
0.1	1.1	1.105	1.105167	1.1051709
0.2	1.2	1.22	1.22133	1.221403
1.0	2	2.5	2.6667	2.718282

The Maclaurin Expansions of Trigonometric Functions

The trigonometric functions sin x, cos x, tan x have derivatives which exist at $x = 0$, and so can be represented by Maclaurin series.

Worked Problem 1.2

Q (a) Find the first three non-zero terms of the Maclaurin series expansion of the sine function.
(b) Deduce the form of the general term, and give the interval of convergence for the series.

A (a) Proceeding in the same manner used for the exponential function:

$f(x) = \sin x$	$f^{(1)}(x) = \cos x$	$f^{(2)}(x) = -\sin x$	$f^{(3)}(x) = -\cos x$	$f^{(4)}(x) = \sin x$	$f^{(5)}(x) = \cos x$
$f(0) = 0$	$f^{(1)}(0) = 1$	$f^{(2)}(0) = 0$	$f^{(3)}(0) = -1$	$f^{(4)}(0) = 0$	$f^{(5)}(0) = 1$

As every other derivative is zero at $x = 0$, we need to go as far as the 5th derivative in order to obtain the first three non-zero terms. Thus, using equation (1.44) we have:

$$f(x) = \sin x = x - \frac{1}{3!}x^3 + \frac{1}{5!}x^5 - \cdots$$

(b) Finding the general term requires some trial and error! In this case:

- The coefficients of even powers of x are zero.
- The denominators in the coefficients of the odd powers of x are odd numbers (formed by adding or subtracting 1 to or from an even number, $2n$).
- As the signs of the coefficients for $c_1, c_3, c_5, \ldots$ alternate, starting with a positive value for c_1, the factor $(-1)^{n-1}$ takes care of the sign alternation.
- Only odd powers of x appear, suggesting that the index can be generated by subtracting 1 from an even number; thus the power of x in the nth term can be written as x^{2n-1} (check that this generates the terms x, x^3 and x^5 by substituting $n = 1$, 2 and 3, respectively). Therefore the general term in this case is given by $\frac{(-1)^{n-1}}{(2n-1)!}x^{2n-1} + \cdots$, where $n = 1, 2, 3, \ldots$

Problem 1.5

Use equation (1.44) to find the first four non-zero terms, as well as the general term, in the Maclaurin series expansions of each of the following functions: (a) e^{-x}; (b) $\cos x$; (c) $(1 - x)^{-1}$.

The Problem with Guessing the General Term: A Chemical Counter Example

In our discussion of the geometric series and the Maclaurin series for sin x, we made the assumption, from the pattern emerging from the first few terms, that we could predict how the series will continue *ad infinitum*. In most cases, this confidence is justified, but sometimes we encounter problems where finding the general term requires a knowledge of the physical context of the problem. An example of such a problem in chemistry involves the computation of the ion–ion interaction energy in an ionic solid, such as NaCl. If we compute the interaction energy arising from the interaction between one ion (positive or negative) and all the other ions in the NaCl lattice structure, then we obtain the Madelung energy in the form:

$$V = -\frac{e^2}{4\pi\varepsilon_0 R}\left\{\frac{6}{\sqrt{1}} - \frac{12}{\sqrt{2}} + \frac{8}{\sqrt{3}} - \frac{6}{\sqrt{4}} + \frac{24}{\sqrt{5}} - \frac{24}{\sqrt{6}} + \cdots\right\} = -\frac{e^2 A}{4\pi\varepsilon_0 R} \quad (1.47)$$

Here A is the Madelung constant for the NaCl structure, and R is the distance between any adjacent Na^+ and Cl^- ions. If we inspect the terms in the series, we can see not only that the sign alternatives but also what appears to be a pattern in the square root values given in the denominators of successive terms. However, in contrast, it is very difficult to see any pattern to the values of the numerators, the reason being that there is none: we can only determine their values from a knowledge of the NaCl structure. In this example, the first term arises from the interaction between a Na^+ ion and the 6 nearest neighbour Cl^- anions at a distance $\sqrt{1}R$; the second term arises from the interaction of nearest-neighbour ions of the same charge, which in this case involves an Na^+ ion and 12 second nearest-neighbour Na^+ ions at a distance $\sqrt{2}R$; the third term is then the interaction between an Na^+ and 8 Cl^- at a distance $\sqrt{3}R$, and so on. The general term in this case is $\frac{m}{R\sqrt{n}}$, where m is the number of nth neighbours at a distance of $R\sqrt{n}$. The next term in the series is, somewhat unexpectedly, $+\frac{12}{\sqrt{8}}$, because the number of 7th nearest neighbours at a distance $R\sqrt{7}$ is zero!

1.4.2 The Taylor Series

Power Series of Functions Expanded About Points Other Than Zero

In many situations, we need to find the power series expansion of a function in terms of the values of the function and its derivatives at some point other than $x = 0$. For example, in the case of a vibrating diatomic molecule, the natural choice of origin for describing the energy of the molecule is the equilibrium internuclear separation, R_e, and not $R = 0$

(where the nuclei have fused!). We can determine the expansion of a function $f(x)$ about an origin $x = a$ (where a is now, by definition, generally not zero) using the Taylor series which is given by the expression:

$$f(x) = f(a) + f^{(1)}(a)(x - a) + \frac{f^{(2)}(a)}{2!}(x - a)^2 + \cdots + \frac{f^{(n)}(a)}{n!}(x - a)^n + \cdots \quad (1.48)$$

Here, $f^{(n)}(a)$ is the value of the nth derivative of $f(x)$ at the point $x = a$. The special case where $a = 0$, as discussed above, generates the Maclaurin series.

Worked Problem 1.3

Q Find the Taylor series expansion for the function $f(x) = e^x$ about the point $x = 1$.

A Since all the derivatives of e^x at $x = 1$ have the value e, the Taylor series takes the form:

$$e^x = e\left\{1 + (x-1) + \frac{(x-1)^2}{2!} + \frac{(x-1)^3}{3!} + \cdots + \frac{(x-1)^{n-1}}{(n-1)!} + \cdots\right\} \quad (1.49)$$

where the last term in the brackets is the nth term. We can see from a plot of the Taylor series expansion of the exponential function, shown in Figure 1.3, that far fewer terms are necessary to achieve a good degree of accuracy in the region around $x = 1$ than is the case with the MacLaurin series. However, we also see that the further away we are from the point $x = 1$, the poorer the approximation and the more terms we will need to achieve a given accuracy. Although it is not obvious from Figure 1.3, the Maclaurin series will be better than this Taylor series expansion for values of x close to $x = 0$.

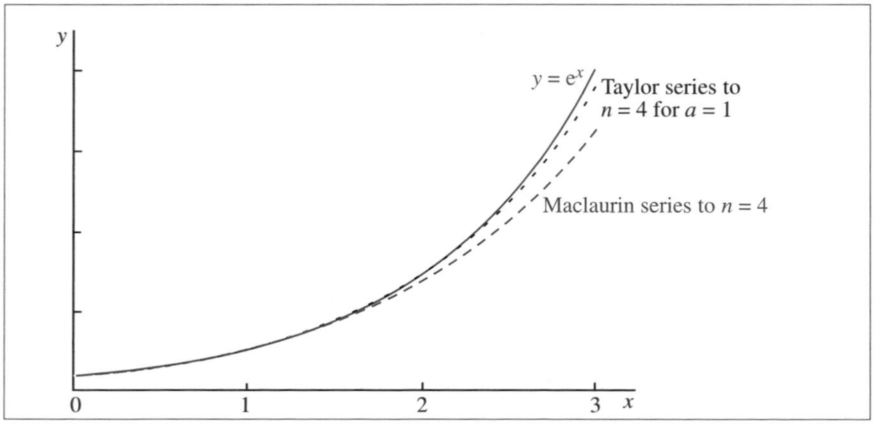

Figure 1.3 An illustration of the improved accuracy achieved with the Taylor series expansion of $f(x)=e^x$ about $x=1$, compared with the Maclaurin series (expanded about $x=0$)

Problem 1.6

Find the first four non-zero terms in the Taylor series expansions of the following functions, expanded about the given point, and deduce the form of the general term for each series: (a) $(1 - x)^{-1}$, at $x = -1$; (b) $\sin x$, at $x = \pi/2$; (c) $\ln x$, at $x = 1$.

Worked Problem 1.4

The variation of potential energy, $E(R)$, with internuclear separation, R, for a diatomic molecule can be approximated by the **Morse potential**, $E(R) = D_e\left\{1 - e^{-\alpha(R-R_e)}\right\}^2$, shown schematically in Figure 1.4. The dissociation energy, D_e and α are both constants for a given molecule.

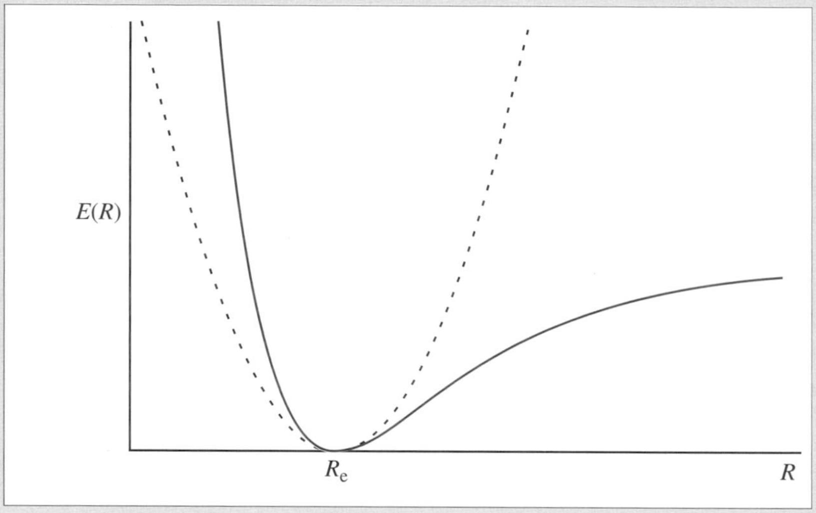

Figure 1.4 Schematic plot of the Morse potential energy function (*full line*). The minimum energy is at $R = R_e$. The harmonic approximation (see text) is shown as a *dashed line*

However, one of the limitations of the Morse potential energy is that, in contrast to the "experimental" curve, the value of the energy at $R = 0$ (corresponding to nuclear fusion) is finite, rather than infinite. The minimum in the Morse potential energy curve occurs at $R = R_e$, which represents the equilibrium bond length.

Q Find the first three terms in the Taylor series expansion for the Morse function about the point $R = R_e$.

A The general expression for the Taylor series expanded about an arbitrary point $x = a$ is:

$$f(x) = f(a) + f^{(1)}(a)(x-a) + \frac{f^{(2)}(a)}{2!}(x-a)^2 + \cdots$$
$$+ \frac{f^{(n)}(a)}{n!}(x-a)^n + \cdots$$

Step 1. Identify R with x, R_e with a, and $E(R)$ with $f(x)$. We can then see that the three terms $f(a)$, $f^{(1)}(a)$ and $f^{(2)}(a)$ are equivalent to $E(R_e)$, $E^{(1)}(R_e)$ and $E^{(2)}(R_e)$, which enables us to re-express the Taylor series in the form:

$$E(R) = E(R_e) + E^{(1)}(R_e)(R-R_e) + \frac{E^{(2)}(R_e)}{2!}(R-R_e)^2 + \cdots \tag{1.50}$$

Thus, we need to evaluate each of the terms $E(R_e)$, $E^{(1)}(R_e)$ and $E^{(2)}(R_e)$, corresponding to the energy and its first two derivatives, evaluated at the point $R = R_e$.
Step 2. Evaluate $E(R_e)$ using the substitution $R = R_e$ in the energy expression:

$$E(R_e) = D_e\left\{1 - e^{-\alpha(R_e - R_e)}\right\}^2 = D_e\{1-1\}^2 = 0$$

since $e^0 = 1$.
Step 3. Evaluate the first and second derivatives, $E^{(1)}(R_e)$ and $E^{(2)}(R_e)$, by applying the chain rule (the detailed working forms part of the next Problem). Thus:

$$E^{(1)}(R) = 2D_e\left\{1 - e^{-\alpha(R-R_e)}\right\} \times \alpha e^{-\alpha(R-R_e)}$$

$$E^{(2)}(R) = 2\alpha^2 D_e\left\{2e^{-2\alpha(R-R_e)} - e^{-\alpha(R-R_e)}\right\}$$

Substituting $R = R_e$ into the above equations gives:

$$E^{(1)}(R_e) = 2D_e\{1-1\} \times \alpha = 0$$

$$E^{(2)}(R_e) = 2\alpha^2 D_e\{2-1\} = 2\alpha^2 D_e$$

Step 4. Substituting the expressions for $E(R_e)$, $E^{(1)}(R_e)$ and $E^{(2)}(R_e)$ into the Taylor expansion (equation 1.50) yields:

$$E(R) = 0 + 0 \times (R - R_e) + \frac{2\alpha^2 D_e}{2!}(R - R_e)^2 + \cdots$$
$$\Rightarrow E(R) = \alpha^2 D_e (R - R_e)^2 + \cdots \qquad (1.51)$$

This is the total energy, $E(R)$, to second order in R, and is commonly known as the **harmonic approximation**. The expression for $E(R)$ gives a good approximation to the potential energy for small displacements of the nuclei, but a somewhat poorer one as the displacements from the equilibrium bond length increase, or decrease, as is seen in Figure 1.4.

Problem 1.7

(a) Use the **chain rule** (see Section 4.2.4 in Volume 1) to find (i) the first and (ii) the second derivative of the Morse function $E(R) = D_e\left\{1 - e^{-\alpha(R-R_e)}\right\}^2$, checking your answers with those given in Worked Problem 1.4.

(b) Verify, by checking the values of $E^{(1)}(R_e)$ and $E^{(2)}(R_e)$, that $E(R_e)$ corresponds to a minimum energy.

(c) Given that the force acting between the nuclei of the molecule is given by $F = -\frac{dE}{dR}$, use equation (1.51) to find an expression for F (for small displacements of the nuclei).

(d) The restoring force acting on a simple harmonic oscillator is given by the expression $F = -kx$. Comment on any similarity between the form of this expression and the one obtained in (c), assuming that the displacement x is equivalent to $(R - R_e)$. What conclusions do you draw about applicability of the harmonic approximation for diatomic molecules?

1.4.3 Manipulating Power Series

Combining Power Series

If two functions are combined by some operation (for example, addition, multiplication, differentiation or integration), then we can find the power series expansion of the resulting function by applying the appropriate operation to the reference series; however, the outcome will be valid only within a domain common to both power series. So, for example, if the Maclaurin series for e^x (interval of convergence: all x in **R**) is multiplied by that for $\ln(1+x)$ (interval of convergence: $-1 < x \leqslant 1$), the resulting series only converges in the common interval of convergence $-1 < x \leqslant 1$.

Worked Problem 1.5

Q Given that the **hyperbolic cosine function** $\cosh x$ is defined by:

$$\cosh x = \frac{1}{2}\{e^x + e^{-x}\}$$

use the Maclaurin series for e^x and e^{-x} to obtain a power series expansion for the $\cosh x$ function. Give the form of the general term.

A If we substitute the Maclaurin series for e^x and e^{-x} in the defining equation for $\cosh x$, we obtain:

$$\cosh x = \frac{1}{2}\left\{1 + x + \frac{x^2}{2!} + \frac{x^3}{3!} + \cdots + \frac{x^{n-1}}{(n-1)!} + \cdots \right.$$
$$\left. + 1 - x + \frac{x^2}{2!} - \frac{x^3}{3!} + \frac{x^4}{4!} - \cdots (-1)^{n-1}\frac{x^{n-1}}{(n-1)!} + \cdots \right\} \quad (1.52)$$
$$\Rightarrow \cosh x = 1 + \frac{x^2}{2!} + \frac{x^4}{4!} + \cdots + \frac{x^{2(r-1)}}{[2(r-1)]!} + \cdots \quad r = 1, 2, 3, \ldots$$

Since both e^x and e^{-x} converge for all x, the above series for $\cosh x$ will also converge for all x.

Problem 1.8

(a) Give the form of the Maclaurin series for the function $\sinh x$, where $\sinh x = \frac{1}{2}(e^x - e^{-x})$.
(b) Deduce the first three terms of the Maclaurin series for the function $f(x) = \dfrac{e^{-x}}{(1-x)}$ using the series for e^{-x} and $\dfrac{1}{(1-x)}$ taken from your answers to Problem 1.5. Give the interval of convergence for $f(x)$.

A Shortcut for Generating Maclaurin Series

Sometimes we can generate Maclaurin series for a given function by a simple substitution. For example, the Maclaurin series for the function e^{-x} can be found as in Problem 1.5; however, an alternative, and much less labour intensive, approach involves writing $X = -x$ and then using the existing series for e^x, with X replacing x.

Worked Problem 1.6

Q Use the substitution $X = -x$ for the Maclaurin series for e^X to find the related series for e^{-x}. How is the interval of convergence for e^{-x} related to that for e^x?

A The Maclaurin expansion for the exponential function e^X is:

$$e^X = 1 + X + \frac{X^2}{2!} + \frac{X^3}{3!} + \frac{X^4}{4!} + \cdots \frac{X^{n-1}}{(n-1)!} + \cdots$$

where X is the independent variable. If we now write $X = -x$ we obtain the series for e^{-x} as required:

$$e^{-x} = 1 - x + \frac{x^2}{2!} - \frac{x^3}{3!} + \frac{x^4}{4!} - \cdots (-1)^{(n-1)} \frac{x^{(n-1)}}{(n-1)!} + \cdots \quad (1.53)$$

The test for absolute convergence shows that the interval of convergence is the same as for e^x.

Problem 1.9

(a) Use the substitution $X = ax$ and the Maclaurin series for e^X to find the series for e^{ax}.

(b) (i) Use the equality $\sin 2x = 2 \sin x \cos x$, and Maclaurin series for $\sin x$ and $\cos x$ to find the first three terms in the related series for $\sin 2x$. (ii) Use the substitution $X = 2x$ and the Maclaurin series for $\sin X$ to find the first three terms in the related series for $\sin 2x$. Compare your answer to (b)(i) above.

1.4.4 The Relationship between Domain and Interval of Convergence

We saw earlier that the Maclaurin series expansion of the function $(1-x)^{-1}$ takes the form $1 + x + x^2 + \cdots$. Although the domain of $f(x)$ includes *all* x values, with the exclusion of $x = 1$, where the function is undefined, the domain of the Maclaurin series, determined by applying the ratio test, is restricted to $-1 < x < 1$. The point $x = 1$ is excluded from the domains of both the function and the series. However, although the point $x = -1$ is clearly included in the domain of the function, since $f(-1) = \frac{1}{2}$, it is *excluded* from the domain of the series. We can further illustrate this by comparing a plot of the function $y = f(x) = (1 - x)^{-1}$ with the MacLaurin series expansion of this function up to the third, fourth, fifth and sixth terms (see Figure 1.5). Clearly the three plots match quite well for $-1 < x < 1$ but differ dramatically for all other values of x. We also see at $x = -1$ that the series representation oscillates between zero and $+1$ as each new term is added to the series, thus indicating divergence at this point.

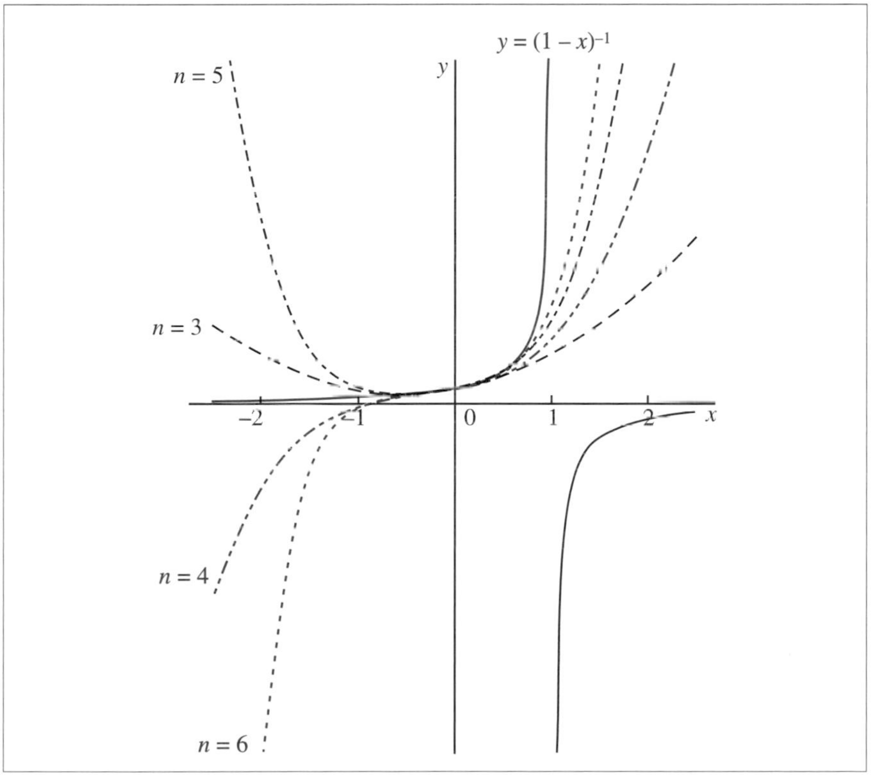

Figure 1.5 A plot of $f(x) = (1-x)^{-1}$ (*full line*), compared with plots of the polynomial truncations of the Maclaurin series expansion $1 + x + x^2 + \cdots + x^{n-1} +$ for $n = 3$–6

1.4.5 Limits Revisited: Limiting Forms of Exponential and Trigonometric Functions

In Chapter 3 of Volume 1 we discussed the behaviour of a function close to some limiting value of the independent variable. Some of the examples concern finite limiting values, but more often we are interested in how functions behave for increasingly small or large values in the independent variable. It is usually straightforward to evaluate the limit of simple functions for increasingly large or small values of x, but for some of the transcendental functions we need to use power series expansions to probe their asymptotic behaviour.

Exponential Functions

The behaviour of the function e^{ax}, as x tends to large or small values, depends upon the signs and magnitudes of x and a. Thus:

- For $x = 0$, $e^{ax} = 1$, irrespective of the value of a.
- For large positive x, e^{ax} increases without limit as x increases for $a > 0$, *but* e^{ax} becomes increasingly small as x increases for $a < 0$.
- Regardless of the signs of x or a, e^{ax} approaches 1, for increasingly small values of x, according to the MacLaurin power series expansion (as seen in Problem 1.9a):

$$e^{ax} = 1 + ax + \frac{(ax)^2}{2!} + \frac{(ax)^3}{3!} + \frac{(ax)^4}{4!} + \cdots \frac{(ax)^{n-1}}{(n-1)!} + \cdots \tag{1.54}$$

Worked Problem 1.7

Q For the radial function of a 3d hydrogen atomic orbital, $R(r) = Nr^2 e^{-r/3a_0}$ (N is a normalizing constant, and a_0 is the Bohr radius), find:
(a) The form of $R(r)$ at small r, using the expansion of the exponential function given above; (b) $\lim_{r\to 0} R(r)$; (c) $\lim_{r\to\infty} R(r)$

A (a) $R(r) \approx Nr^2(1 - \frac{r}{3a_0} + \cdots) = Nr^2$ for small r.
(b) Using the approximation from (a), we see that $\lim_{r\to 0} R(r) = 0$
(c) For large r we see that the limiting value of the function will be determined by the outcome of the competition between the Nr^2 term and the $e^{-r/3a_0}$ term. As we saw in Section 2.3.4 in Volume 1, the exponential term will always overcome the power term, and so $\lim_{r\to\infty} R(r) = 0$.

Problem 1.10

The Einstein model for the molar heat capacity at constant volume, C_V, of a solid yields the formula:

$$C_V = 3R\left(\frac{h\nu}{kT}\right)^2 \left\{\frac{e^{h\nu/2kT}}{e^{h\nu/2kT} - 1}\right\}^2$$

Show that at high values of T, when we can justifiably substitute the exponential terms by their two-term series approximations, C_V tends to the limit $3R$.

Trigonometric Functions

In an analogous way, the series expansions for the sine and cosine functions have the forms:

$$\cos ax = 1 - \frac{(ax)^2}{2!} + \cdots; \quad \sin ax = ax - \frac{(ax)^3}{3!} + \cdots \tag{1.55}$$

as $x \to 0$. For very small values of x, cos ax and sin ax may be approximated by 1 and ax, respectively. However, as x increases without limit, in both positive and negative senses, the values of the sine or cosine functions oscillate between ± 1.

Problem 1.11

Consider a particle confined to move in a constant potential between the points $x = 0$ and $x = L$ at which the potential is infinite. The associated wavefunction has the form:

$$\psi = \sqrt{\frac{2}{L}} \sin \frac{n\pi x}{L}$$

where n is the quantum number defining the state of the particle, and has values 1, 2, 3, ... Find the expression for ψ: (a) at $x = 0$; (b) at $x = L$; (c) when x is very small.

Summary of Key Points

The key points discussed in this chapter include:

1. The definition of finite sequences with examples including the geometric and arithmetic progressions.

2. The definition of indefinite sequences and the concepts of convergence and divergence of a sequence of numbers.

3. The distinction between a finite series having a finite sum and an infinite series where a finite sum exists only if the series converges.

4. Testing an infinite series for convergence: the ratio test for absolute convergence and the interval of convergence.

5. Power series: the Maclaurin and Taylor series.

6. Power series expansions of functions: the appropriate choice of expansion point.

7. Truncation of power series.

8. Determining the general term in a power series.

9. The Taylor series expansion of the Morse potential leading to the harmonic approximation.

10. Manipulating power series.

11. Using power series expansions of functions to probe limiting behaviour for increasingly large or small values of the independent variable.

2
Numbers Revisited: Complex Numbers

In Chapter 2 of Volume 1 we saw that the solution of a **quadratic equation** of the form:

$$ax^2 + bx + c = 0 \tag{2.1}$$

can yield up to two real roots, depending on the values of the coefficients, a, b and c. The general solution to quadratic equations of this form is given by the formula:

$$x = \frac{-b \pm \sqrt{b^2 - 4ac}}{2a} \tag{2.2}$$

where the quantity $b^2 - 4ac$ is known as the **discriminant** (see Section 2.4 in Volume 1). If the discriminant is positive, then the equation has two real and different roots; if it is zero, then the equation will have two identical roots; and if it is negative, there are no real roots, as the formula involves the square root of a negative number. For example, the equation:

$$x^2 - 4x + 3 = 0$$

yields two real roots, $x = 3$ and $x = 1$, according to:

$$x = \frac{4 \pm \sqrt{16 - (4 \times 3)}}{2} = 2 \pm \frac{\sqrt{4}}{2} = 2 \pm 1 = 3, 1$$

We can represent this solution graphically (see Figure 2.1) in terms of where the function $y = x^2 - 4x + 3$ cuts the x-axis, where $y = 0$.

However, if we use equation (2.2) to find the roots of the quadratic equation:

$$x^2 - 4x + 6 = 0$$

we find that the solution yields:

$$x = \frac{4 \pm \sqrt{16 - (4 \times 6)}}{2} = 2 \pm \frac{\sqrt{-8}}{2} \tag{2.3}$$

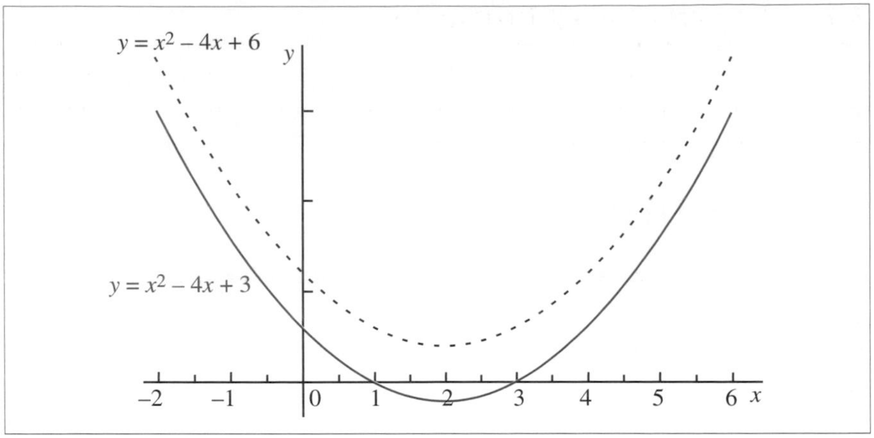

Figure 2.1 Plots of the two functions $y = x^2 - 4x + 3$ and $y = x^2 - 4x + 6$, showing the presence of two and zero real roots of the respective equations $x^2 - 4x + 3 = 0$ and $x^2 - 4x + 6 = 0$

which requires us to find the square root of -8. Graphically, we see in Figure 2.1 that a plot of the function $y = x^2 - 4x + 6$ does not cut the x-axis at all. Logic would seem to dictate that any solution to the second of these two equations is nonsensical, and that the result cannot possibly be real – especially when we view the plot of the function, which clearly does not cut the x-axis! However, there is a way of circumventing this problem by simply extending the number system to include so-called **complex numbers**, which incorporate $\sqrt{-1}$ as a legitimate number. This concept can naturally seem somewhat bemusing but, once we get over the shock, we find that the treatment of complex numbers is really quite straightforward and, more importantly, we find that they allow us to tackle real problems in chemistry in a way that would otherwise be impossible.

Aims

This chapter extends the familiar number system to include complex numbers containing the imaginary number i. By the end of this chapter, you should be able to:

- Recognize the real and imaginary parts of a complex number expressed in either cartesian or plane polar coordinates
- Determine the modulus and argument of a complex number, and denote its location on an Argand diagram
- Perform arithmetical operations on complex numbers
- Use the Euler formula and the De Moivre theorem to evaluate powers of complex numbers, to determine nth roots of a complex number, and to identify real and imaginary parts of functions of a complex variable

2.1 The Imaginary Number i

As we saw above, the solutions to algebraic equations do not always yield real numbers. For example, the solution of the equation $x^2 + 1 = 0$ yields the apparently meaningless result:

$$x = \pm\sqrt{-1} \tag{2.4}$$

because the square root of a negative number is not defined in terms of a real number. However, if we now *define* the **imaginary number** $\mathrm{i} = \sqrt{-1}$, then the two roots may be specified as $x = \pm\mathrm{i}$. In general, an imaginary number is defined as any real number multiplied by i. Thus, for example, the number $\sqrt{-8}$, which emerged from the solution to equation (2.3) above, can be written as $\sqrt{8}\sqrt{-1} = \sqrt{8}\mathrm{i}$.

Worked Problem 2.1

Q Solve the quadratic equation $x^2 + 2x + 5 = 0$.

A The formula given in equation (2.2) for the roots of a quadratic equation yields:

$$x = -1 \pm \frac{\sqrt{4-20}}{2} = -1 \pm \frac{\sqrt{-16}}{2} = -1 \pm \frac{4}{2}\sqrt{-1} = -1 \pm 2\mathrm{i}$$

Problem 2.1

(a) Draw plots of the following functions: (i) $y = x^2 - 2x - 3$ and (ii) $y = x^2 - 2x + 2$. In each case, comment on whether the plot cuts the x-axis and, if so, where?

(b) Use equation (2.2) to find the roots of each of the quadratic equations $x^2 - 2x - 3 = 0$ and $x^2 - 2x + 2 = 0$. Comment on your answers with respect to your plots from part (a).

2.2 The General Form of Complex Numbers

In the answer to Worked Problem 2.1, we obtained the required roots of the quadratic equation in the form of a sum of a real number (–1) and an imaginary number (2i or –2i). Such numbers are termed **complex numbers**, and have the general form:

$$z = x + \mathrm{i}y \tag{2.5}$$

The net result is that the original quotient is transformed into the form of a complex number with real and imaginary parts, $\frac{x_1x_2 + y_1y_2}{x_2^2 + y_2^2}$ and $\frac{y_1x_2 - x_1y_2}{x_2^2 + y_2^2}$, respectively.

Worked Problem 2.3

Q Using z_1 and z_2 as defined in Worked Problem 2.2, express z_1/z_2 in the form $x + iy$.

A $\frac{z_1}{z_2} = \frac{1+2i}{-2+i} = \frac{1+2i}{-2+i} \times \frac{-2-i}{-2-i} = \frac{-2-i-4i+2}{4+2i-2i+1} = \frac{-5i}{5} = -i.$
In this example, the answer is an imaginary number, since $x = 0$ and $y = -1$.

Problem 2.5

Express the following in the form $x + iy$: (a) $\frac{1}{i}$; (b) $\frac{1-i}{2-i}$; (c) $\frac{i(2+i)}{(1-2i)(2-i)}$

2.4 The Argand Diagram

Since a complex number is defined in terms of two real numbers, it is convenient to use a graphical representation in which the real and imaginary parts define a point (x,y) in a plane. Such a representation is provided by an **Argand diagram**, as seen in Figure 2.2.

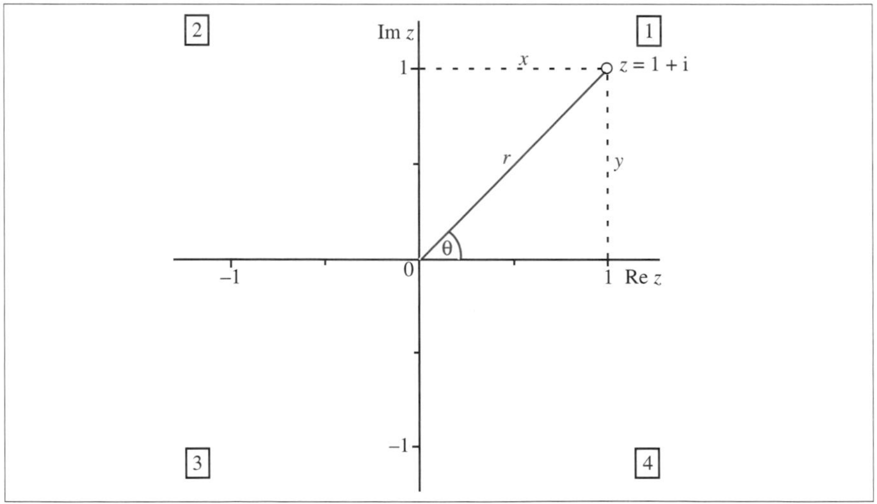

Figure 2.2 An Argand diagram displaying the complex number $z = 1+i$, in terms of the Cartesian coordinates (1,1) or, alternatively, in terms of the polar coordinates ($r = \sqrt{2}, \theta = \pi/4$)

Problem 2.6

Plot your answers to Problem 2.3 as points in an Argand diagram.

The location of z in the Argand diagram can be specified by using either **Cartesian coordinates** (x,y), where $x = \text{Re } z$, $y = \text{Im } z$, or **polar coordinates** (r,θ), where $r \geqslant 0$ and $-\pi < \theta \leqslant \pi$. The reason for choosing this range of θ values, rather than $0 < \theta \leqslant 2\pi$, derives from the convention that θ should be positive in the first two quadrants (above the x-axis), moving in an anticlockwise sense from the Re z axis, and negative in the third and fourth quadrants (below the x-axis), moving in a clockwise sense. The quadrant numbering runs from 1 to 4, in an anticlockwise direction, as indicated in Figure 2.2.

2.4.1 The Modulus and Argument of *z*

The polar coordinates r and θ define the **modulus** (alternatively known as the **absolute value** and sometimes denoted by $|z|$) and **argument**, respectively, of z. From Pythagoras' theorem, and simple trigonometry, the modulus and argument of z are defined as follows (see Figure 2.2):

$$r = \sqrt{x^2 + y^2}, \quad r \geqslant 0 \tag{2.11}$$

$$\tan\theta = \frac{y}{x} \Rightarrow \theta = \tan^{-1}(y/x) \tag{2.12}$$

Great care is required in determining θ, because it is easy to make a mistake in specifying the correct quadrant. For example, although the complex numbers $z = 1 - \text{i}$ and $z = -1 + \text{i}$ both have $\tan\theta = -1$, they lie in the fourth and second quadrants, respectively, as seen in Figure 2.3.

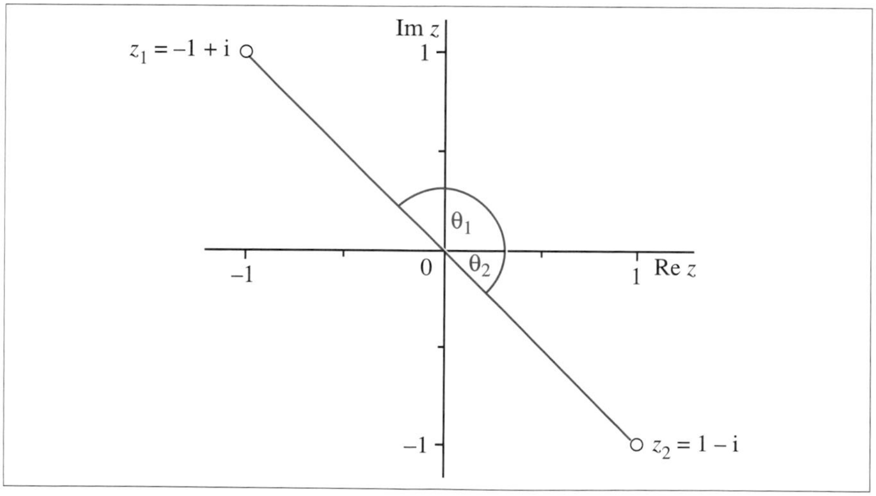

Figure 2.3 An Argand diagram showing the complex numbers $z_1 = -1 + \text{i}$ and $z_2 = 1 - \text{i}$ with modulus $\sqrt{2}$ and arguments $3\pi/4$ and $-\pi/4$, respectively

2.5.1 Euler's Formula

If we equate equation (2.18) with equation (2.13) we obtain:

$$z = re^{i\theta} = r(\cos\theta + i\sin\theta) \qquad (2.19)$$

which on cancelling r yields:

$$e^{i\theta} = \cos\theta + i\sin\theta \qquad (2.20)$$

This important result is known as **Euler's formula**.

Worked Problem 2.5

Q Express the complex conjugate of $z = r(\cos\theta + i\sin\theta)$ in polar form.

A The complex conjugate of z can be written in terms of r,θ as above, using the Maclaurin series for $\cos\theta$ and $\sin\theta$, as:

$$z^* = r(\cos\theta - i\sin\theta) = r\left\{1 - i\theta - \frac{\theta^2}{2!} + \frac{i\theta^3}{3!} + \frac{\theta^4}{4!} - \frac{i\theta^5}{5!} + \cdots\right\}$$

which can be rewritten as:

$$z^* = r\left\{1 - i\theta + \frac{(i\theta)^2}{2!} - \frac{(i\theta)^3}{3!} + \frac{(i\theta)^4}{4!} - \frac{(i\theta)^5}{5!} + \cdots\right\}$$

The part in braces is the Maclaurin series for $e^{-i\theta}$ and so we can now express z^* as:

$$z^* = re^{-i\theta} \qquad (2.21)$$

The Number $e^{i\pi}$

Using Euler's formula to evaluate $e^{i\pi}$ we see that:

$$e^{i\pi} = \cos\pi + i\sin\pi \qquad (2.22)$$

However, as $\cos\pi = -1$ and $\sin\pi = 0$, we obtain the extraordinary and elegant result that:

$$e^{i\pi} = -1 \qquad (2.23)$$

which rearranges to a single relationship:

$$e^{i\pi} + 1 = 0 \qquad (2.24)$$

containing the irrational numbers e and π, the imaginary number i, as well as the numbers zero and unity.

2.5.2 Powers of Complex Numbers

The advantage of using the polar form for z is that it makes certain manipulations much easier. Thus, for example, we can obtain:

- The modulus of z directly from the product of z and z^*, using equations (2.18) and (2.21):

$$zz^* = r^2 e^{i\theta} e^{-i\theta} = r^2 \Rightarrow r = \sqrt{zz^*} \quad (2.25)$$

- Positive and negative powers of z:

$$z^n = (re^{i\theta})^n = r^n e^{in\theta} \quad (n = \pm 1, \pm 2, \pm 3, \pm 4, \ldots) \quad (2.26)$$

  where, for a given value of n, z^n is seen to be a complex number, with modulus r^n and argument $n\theta$.
- Rational powers of z, where $n = p/q$ $(q \neq 0)$:

$$z^{p/q} = r^{p/q} e^{i(p/q)\theta} \quad (2.27)$$

Worked Problem 2.6

Q If $z = \cos\theta + i\sin\theta$, show that $1/z$ is the complex conjugate of z.

A As z has a unit modulus ($r = 1$), $z = e^{i\theta}$, and $1/z = \frac{1}{e^{i\theta}} = e^{-i\theta}$, which is the complex conjugate of z (see equation 2.21).

Problem 2.9

For the two complex numbers $z_1 = r_1 e^{i\theta_1}$ and $z_2 = r_2 e^{i\theta_2}$, give expressions for the modulus and argument of: (a) $z_1 z_2$; (b) z_1/z_2; (c) ${z_1}^2/{z_2}^4$.

Problem 2.10

Express $z = -1 - i$ in polar form, and thus determine the modulus and argument (in radians) of z^2 and z^{-4}.

2.5.3 The De Moivre Theorem

We have seen from equation (2.26) that the nth power of a complex number can be expressed as:

$$z^n = r^n e^{in\theta} \quad (2.28)$$

with modulus and argument r^n and $n\theta$, respectively. Using Euler's formula, equation (2.28) becomes:

$$z^n = r^n e^{in\theta} = r^n(\cos n\theta + i \sin n\theta) = r^n(\cos\theta + i\sin\theta)^n \tag{2.29}$$

After cancelling the r^n factors in equation (2.29), we obtain the **De Moivre theorem**:

$$(\cos\theta + i\sin\theta)^n = \cos n\theta + i\sin n\theta \tag{2.30}$$

Problem 2.11

(a) Show that $\dfrac{1}{(\cos\theta + i\sin\theta)} = \cos\theta - i\sin\theta$.
(b) Give an expression for $(\cos\theta + i\sin\theta)^{1/2}$.
(c) Use equation (2.29) to give expressions for the real and imaginary parts of z^n.
(d) Find the real and imaginary parts of z^3 and z^{-2}, where $z = -1 - i$.

2.5.4 Complex Functions

So far we have been concerned largely with the concept of the complex number, but we can see from our discussion of Euler's formula that the general form of a complex number actually represents a complex mathematical function, say $f(\theta)$, where:

$$f(\theta) = \cos\theta + i\sin\theta \tag{2.31}$$

This function comprises a real part and an imaginary part, and so in general we can define a complex function in the form:

$$f(x) = g(x) + ih(x) \tag{2.32}$$

where the complex conjugate of the function is given by:

$$f(x)^* = g(x) - ih(x) \tag{2.33}$$

Thus $f(x)f(x)^*$ is a real function of the form:

$$f(x)f(x)^* = g(x)^2 + h(x)^2 \tag{2.34}$$

The property of complex functions given in equation (2.34) plays a very important role in quantum mechanics, where the wave function of an electron, ψ, which may be complex in form, is related to the physically meaningful probability density through the product $\psi\psi^*$. If ψ is a complex function, then, from equation (2.34), $\psi\psi^*$ is a real function.

The Periodicity of the Exponential Function

It may seem odd to think of the exponential function, $z = e^{i\theta}$, as periodic because it is clearly not so when the exponent is real. However, the presence of the imaginary number i in the exponent allows us to define a modulus and argument as 1 and θ, respectively. If we represent the values of the function on an Argand diagram, we see that they lie on a circle of radius, $r = 1$, in the complex plane (see Figure 2.4). Different values of θ then define the location of complex numbers of modulus unity on the circumference of the circle. We can also see that the function is periodic, with period 2π:

$$e^{i(\theta+2m\pi)} = e^{i\theta} \times e^{i2m\pi} = e^{i\theta} \tag{2.35}$$

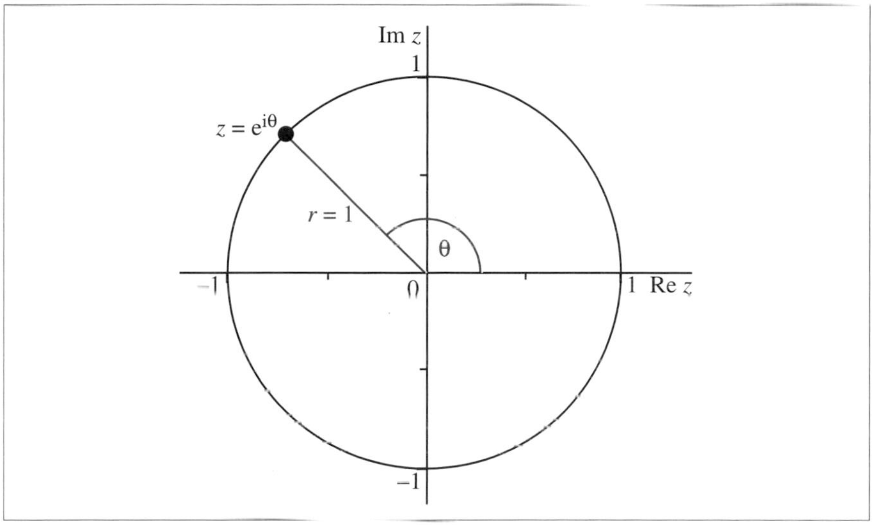

Figure 2.4 The function $z=e^{i\theta}$ is periodic in the complex plane, with period 2π

> **Problem 2.12**
>
> Use Euler's formula to show that $e^{i2\pi} = 1$, and hence prove that $e^{i(\theta+2m\pi)} = e^{i\theta}$, for $m = 1, 2, ...$

> **Problem 2.13**
>
> (a) Use De Moivre's theorem to show that $e^{-i\theta} = \cos\theta - i\sin\theta$.
> (b) Use Euler's formula, and the result given in part (a), to show that:
> (i) $\cos\theta = \frac{1}{2}\left(e^{i\theta} + e^{-i\theta}\right)$ and (ii) $\sin\theta = \frac{1}{2i}\left(e^{i\theta} - e^{-i\theta}\right)$.

Problem 2.14

The solution of the differential equation describing the simple harmonic oscillator problem (see Worked Problem 7.4(c) in Volume 1) is:

$$y = A\cos kt + B\sin kt$$

Derive an alternative form for the solution, using the function e^{ikt}, and its complex conjugate, e^{-ikt}.
Hint: use the results given in Problem 2.13(b).

The Eigenvalue Problem Revisited

The three 2p orbitals resulting from the solution of the **Schrödinger equation** for the hydrogen atom can be written as:

$$\psi_1 = N_1 e^{-r/2a_0} r\sin\theta\, e^{i\phi};\quad \psi_0 = N_2 e^{-r/2a_0} r\cos\theta;\quad \psi_{-1} = N_1 e^{-r/2a_0} r\sin\theta\, e^{-i\phi}$$

where N_1 and N_2 are constants, a_0 is the Bohr radius, r is the distance of the electron from the nucleus, and the suffix attached to each ψ indicates the value of the orientation quantum number m_l. The Schrödinger equation, $\hat{H}\psi = E\psi$, is an example of an **eigenvalue problem** (see Sections 4.3.1 and 7.4.3 in Volume 1) where, in this case, $\hat{H}$ is an operator known as the **Hamiltonian**; E is the **eigenvalue** (corresponding to the energy of the system) and ψ is the **eigenfunction** (or wave function). As we saw in Volume 1, if two functions are both solutions to an eigenvalue problem, and associated with the same eigenvalue, then a linear combination of the functions will also be a solution. We can use this property to construct real orbital functions that we can visualize more easily. We explore this idea a little further in the next problem.

Problem 2.15

(a) Find the real and imaginary parts of each of the three 2p orbitals given above.
(b) Use the results given in Problem 2.13(b) to show that the following linear combinations yield real functions:

(i) $$\frac{1}{\sqrt{2}}(\psi_1 + \psi_{-1}) = \sqrt{2}N_1 e^{-r/2a_0} r\sin\theta\cos\phi$$

(ii) $$-\frac{i}{\sqrt{2}}(\psi_1 - \psi_{-1}) = \sqrt{2}N_1 e^{-r/2a_0} r\sin\theta\sin\phi$$

(c) Given that $x = r\sin\theta\cos\phi$, $y = r\sin\theta\sin\phi$ and $z = r\cos\theta$, rewrite the three *real* atomic orbital functions, ψ_0, $\frac{1}{\sqrt{2}}(\psi_1 + \psi_{-1})$ and $-\frac{i}{\sqrt{2}}(\psi_1 - \psi_{-1})$, in terms of the independent variables x, y and z and hence propose new labels for the three wavefunctions.

Structure Factors in Crystallography

The intensity of the scattered beam of X-rays from the (hkl) plane of a crystal is proportional to FF^*, where F, the structure factor, is given by:

$$F(hkl) = \sum_{j}^{\text{cell}} f_j \, e^{2\pi i[hx_j + ky_j + lz_j]} \tag{2.36}$$

The summation runs over the appropriate number of atoms in the unit cell with (fractional) coordinates (x_j,y_j,z_j) and scattering factor f_j.

Problem 2.16

Metallic sodium has a body-centred cubic structure with two atoms per unit cell located at (0,0,0) and $(\frac{1}{2}, \frac{1}{2}, \frac{1}{2})$, respectively.

(a) Use equation (2.36) to show that $F(hkl) = f_{\text{Na}} + f_{\text{Na}} e^{\pi i(h+k+l)}$ and, with the aid of Euler's formula, determine its real and imaginary parts.
(b) Show that reflections occur [*i.e.* when $F(hkl) \neq 0$] only if $h + k + l$ is even.

2.5.5 Roots of Complex Numbers

The polar form of a complex number, z, raised to the power n is given in equation (2.28) as:

$$z^n = (re^{i\theta})^n = r^n e^{in\theta} \tag{2.37}$$

De Moivre's theorem allows us to express z^n in the form:

$$z^n = r^n(\cos n\theta + i\sin n\theta) \tag{2.38}$$

It follows that one square root of a complex number (where $n = \frac{1}{2}$) is given by:

$$z^{1/2} = r^{1/2}\left(\cos\frac{\theta}{2} + i\sin\frac{\theta}{2}\right) \tag{2.39}$$

The method used to retrieve the second square root is now described in Worked Problem 2.7.

Worked Problem 2.7

Q Use equation (2.37) and the periodicity of $e^{i\theta}$ (see Problem 2.12) to find the two square roots of –1.

A Substituting equation (2.35) for $e^{i\theta}$ in equation (2.37) yields:

$$z^n = \left(re^{i(\theta+2m\pi)}\right)^n = r^n e^{i(\theta+2m\pi)n}, \quad m = 1, 2, 3, \ldots$$

Now, the number (-1) has $r = 1$ and $\theta = \pi$; *hence:*

$$(-1)^{1/2} = e^{i(\pi+2m\pi)\times 1/2}, m = 1, 2, 3, \ldots$$

$$= e^{i(\pi/2+m\pi)}, m = 1, 2, 3, \ldots$$

$$= \cos\left(\frac{\pi}{2} + m\pi\right) + i\sin\left(\frac{\pi}{2} + m\pi\right), m = 1, 2, 3, \ldots$$

Thus for $m = 1$:

$$z^{1/2} = \cos\frac{3\pi}{2} + i\sin\frac{3\pi}{2} = -i$$

For $m = 2$:

$$z^{1/2} = \cos\frac{5\pi}{2} + i\sin\frac{5\pi}{2} = i$$

For $m = 3$:

$$z^{1/2} = \cos\frac{7\pi}{2} + i\sin\frac{7\pi}{2} = -i$$

For $m = 4$:

$$z^{1/2} = \cos\frac{9\pi}{2} + i\sin\frac{9\pi}{2} = i$$

and so on. We see that taking $m \geqslant 3$ merely replicates the roots already found, and so the two square roots of -1 are $\pm i$.

This method can be extended to find the nth roots of any number.

Problem 2.17

Show that the three cube roots of i (given by $i^{1/3}$) are $-\frac{\sqrt{3}}{2} + \frac{1}{2}i$, $-i$ and $\frac{\sqrt{3}}{2} + \frac{1}{2}i$.

Summary of Key Points

This chapter introduces imaginary and complex numbers as a legitimate extension of the number system. The key points discussed in this chapter include:

1. An introduction of the imaginary number $\mathrm{i} = \sqrt{-1}$ as a means to finding all roots of polynomial equations.

2. A definition of the general form of a complex number, $z = x + iy$, comprising real and imaginary parts.

3. The algebra of complex numbers: addition, subtraction and multiplication.

4. The complex conjugate and division of complex numbers.

5. The graphical representation of the complex number through the Argand diagram.

6. The definition of modulus and argument of a complex number.

7. The polar form of complex numbers, $z = r\mathrm{e}^{\mathrm{i}\theta}$.

8. Euler's formula, $\mathrm{e}^{\mathrm{i}\theta} = \cos\theta + \mathrm{i}\sin\theta$.

9. Powers of complex numbers and de Moivre's theorem, $\mathrm{e}^{in\theta} = \cos n\theta + \mathrm{i}\ \sin n\theta$.

10. Complex functions.

11. The periodicity of the exponential function, $\mathrm{e}^{\mathrm{i}\theta}$, and the modelling of wave phenomena.

12. The real and complex forms of atomic orbitals.

13. Finding the roots of positive, negative and complex numbers.

3
Working with Arrays I: Determinants

In all areas of the physical sciences we encounter problems that require the solution of sets of simultaneous **linear equations**. These range from seemingly mundane everyday problems to highly complex problems in quantum mechanics or spectroscopy requiring the solution of hundreds of simultaneous linear equations. For small numbers of such equations, the solutions may be most straightforwardly obtained using the methods of elementary algebra. However, as the number of equations increases, their alegbraic solution becomes cumbersome and ultimately intractable. In this chapter we introduce the concept of the determinant to provide one of the tools used to solve problems involving large numbers of simultaneous equations. The other tools required for solving systems of linear equations are provided by matrix algebra, which we discuss in detail in Chapter 4.

Aims

This chapter introduces the determinant as a mathematical object which we can use to tackle problems involving large numbers of simultaneous linear equations. By the end of this chapter, you should be able to:

- Appreciate that a determinant expands to yield an expression or value
- Recognize how determinants can be used to solve simultaneous linear equations
- Expand determinants of second and third order about a given row or column
- Use the properties of determinants to introduce as many zeros as possible to the right (or left) of the leading diagonal
- Define and evaluate the first-order cofactors of a determinant

3.1 Origins: the Solution of Simultaneous Linear Equations

We begin our discussion of linear systems by introducing the determinant as a tool for solving sets of simultaneous linear equations in which the indices of the unknown variables are all unity. Consider the pair of equations:

$$a_{11}x + a_{12}y = b_1 \tag{3.1}$$

$$a_{21}x + a_{22}y = b_2 \tag{3.2}$$

where a_{11}, a_{12}, a_{21}, a_{22}, b_1 and b_2 are constant coefficients and x and y are the "unknowns". We can determine the unknowns using elementary algebra as follows:

- Multiply equation (3.1) by a_{22} and equation (3.2) by a_{12} to give:

$$a_{11}a_{22}x + a_{12}a_{22}y = b_1a_{22} \tag{3.3}$$

$$a_{12}a_{21}x + a_{12}a_{22}y = b_2a_{12} \tag{3.4}$$

- Subtract equation (3.4) from equation (3.3) to yield:

$$(a_{11}a_{22} - a_{12}a_{21})x = b_1a_{22} - b_2a_{12}$$

which we can rearrange to give an expression for x in terms of the constant coefficients:

$$x = \frac{b_1a_{22} - b_2a_{12}}{a_{11}a_{22} - a_{12}a_{21}} \tag{3.5}$$

- Now, multiply equation (3.1) by a_{21} and equation (3.2) by a_{11} and subtract the resulting equations to yield:

$$y = \frac{b_2a_{11} - b_1a_{21}}{a_{11}a_{22} - a_{12}a_{21}} \tag{3.6}$$

- The denominators in equations (3.5) and (3.6) are the same, and can be written alternatively as:

$$\begin{vmatrix} a_{11} & a_{12} \\ a_{21} & a_{22} \end{vmatrix} = a_{11}a_{22} - a_{12}a_{21} \tag{3.7}$$

The symbol on the left side of equation (3.7) defines a **determinant** of order 2, the expansion of which is given on the right. We can similarly express the numerators as determinants of order 2, and we see that the value of the two unknowns is then given by the ratio of two determinants:

$$x = \frac{\begin{vmatrix} b_1 & b_2 \\ a_{12} & a_{22} \end{vmatrix}}{\begin{vmatrix} a_{11} & a_{12} \\ a_{21} & a_{22} \end{vmatrix}} \quad \text{and} \quad y = \frac{\begin{vmatrix} b_2 & b_1 \\ a_{21} & a_{11} \end{vmatrix}}{\begin{vmatrix} a_{11} & a_{12} \\ a_{21} & a_{22} \end{vmatrix}} \tag{3.8}$$

The purpose of introducing this notation is that it readily extends to n linear algebraic equations in n unknowns. The problem then reduces to one of evaluating the respective determinants of order n.

Problem 3.1

Solve the following simultaneous equations for x and y by evaluating the appropriate determinants according to equation (3.8):

$$2x + y = 5 \tag{3.9}$$

$$\tfrac{1}{2}x + 8y = 9 \tag{3.10}$$

Hint: you will need to associate each of the coefficients a_{11}, a_{12}, a_{21}, a_{22}, b_1 and b_2 in equations (3.1) and (3.2) with those in equations (3.9) and (3.10).

This type of problem arises in a chemical context quite frequently. For example, the activation energy of a chemical reaction can be determined by measuring the rate constant for a particular reaction at two different temperatures. The relationship between rate constant and temperature is given by the **Arrhenius equation**:

$$k = A\mathrm{e}^{-E_\mathrm{a}/RT} \tag{3.11}$$

where E_a is the activation energy for the reaction, and A is the so-called pre-exponential factor. We can convert the Arrhenius equation to a linear form by taking logs of both sides:

$$\ln k = \ln A - \frac{E_\mathrm{a}}{RT} \tag{3.12}$$

If we now measure the rate constant at two different temperatures, T_1 and T_2, we obtain a pair of simultaneous linear equations which we can solve for the two unknowns, E_a and $\ln A$:

$$\ln k_1 = \ln A - \frac{E_\mathrm{a}}{RT_1} \tag{3.13}$$

$$\ln k_2 = \ln A - \frac{E_\mathrm{a}}{RT_2} \tag{3.14}$$

Problem 3.2

(a) By analogy with equations (3.1) and (3.2), identify which terms in equations (3.13) and (3.14) correlate with the constant coefficients a_{11}, a_{12}, a_{21}, a_{22}, b_1 and b_2 and which terms correlate with the unknowns, x and y.

(b) If the temperatures on the Fahrenheit and Centigrade scales are T and t, respectively, we can express the relationship between the two as $T = at + b$, where a and b are unknowns. By analogy with equations (3.1) and (3.2), use the method of determinants to obtain the values of a and b by using the boiling and freezing points of water on the two temperature scales. Hence find the formula relating T to t.

Hint: the boiling and freezing points of water on the Fahrenheit scale are $T_b = 212\ °F$ and $T_f = 32\ °F$.

3.2 Expanding Determinants

In general, a determinant of **order** n is defined as a square array of n^2 elements arranged in n rows and n columns:

$$\begin{vmatrix} a_{11} & a_{12} & \cdots & a_{1n} \\ a_{21} & a_{22} & \cdots & a_{2n} \\ \vdots & \vdots & \ddots & \vdots \\ a_{n1} & a_{n2} & \cdots & a_{nn} \end{vmatrix} \tag{3.15}$$

The **elements** of this determinant are denoted by a_{ij} or b_{ij}, where i denotes the row and j the column number. Note that the letter used commonly derives from the label applied to a related square matrix – a consequence of the common definition of a determinant as an operation on a square matrix (see Section 4.1 in Chapter 4). We have seen above in equation (3.7) that a determinant of order 2 is evaluated in terms of the elements a_{ij} which lie at the intersection of the ith row with the jth column of the determinant.

A determinant of order 3, which might result from a problem involving three simultaneous equations in three unknowns, is expanded as follows:

$$\begin{vmatrix} a_{11} & a_{12} & a_{13} \\ a_{21} & a_{22} & a_{23} \\ a_{31} & a_{32} & a_{33} \end{vmatrix} = a_{11}\begin{vmatrix} a_{22} & a_{23} \\ a_{32} & a_{33} \end{vmatrix} - a_{12}\begin{vmatrix} a_{21} & a_{23} \\ a_{31} & a_{33} \end{vmatrix} + a_{13}\begin{vmatrix} a_{21} & a_{22} \\ a_{31} & a_{32} \end{vmatrix} \tag{3.16}$$

This **expansion** proceeds by taking the elements of the first row in turn, and multiplying each one by the determinant of what remains on crossing out the row and column containing the chosen element, and then attaching the sign $(-1)^{i+j}$. The signed determinants of order 2 in equation (3.16) are known as the first-order **cofactors** A_{11}, A_{12} and A_{13} of the three elements a_{11}, a_{12} and a_{13}, respectively.

In general, the n^2 cofactors of any determinant of order n are obtained by deleting *one* row and *one* column to form a determinant of order $n - 1$, the value of which is multiplied by an odd or even power of -1, depending upon the choice of row index and column index. Thus, if the ith row and jth column of a determinant of order n are both deleted, then the ijth cofactor, A_{ij}, is formed from the value of the resulting determinant of order $n - 1$, multiplied by $(-1)^{i+j}$. For example, the cofactor A_{12} of the determinant of order 3 in equation (3.16) is obtained by deleting the *first* row and the *second* column of the determinant, and multiplying the resulting determinant of order 2 by $(-1)^{1+2}$:

$$A_{12} = (-1)^{1+2}\begin{vmatrix} a_{21} & a_{23} \\ a_{31} & a_{33} \end{vmatrix} = -\begin{vmatrix} a_{21} & a_{23} \\ a_{31} & a_{33} \end{vmatrix} \tag{3.17}$$

Rewriting equation (3.16) in terms of the three cofactors:

$$A_{11} = \begin{vmatrix} a_{22} & a_{23} \\ a_{32} & a_{33} \end{vmatrix}, A_{12} = -\begin{vmatrix} a_{21} & a_{23} \\ a_{31} & a_{33} \end{vmatrix} \quad \text{and} \quad A_{13} = \begin{vmatrix} a_{21} & a_{22} \\ a_{31} & a_{32} \end{vmatrix} \tag{3.18}$$

yields:

$$\begin{vmatrix} a_{11} & a_{12} & a_{13} \\ a_{21} & a_{22} & a_{23} \\ a_{31} & a_{32} & a_{33} \end{vmatrix} = a_{11}A_{11} + a_{12}A_{12} + a_{13}A_{13} \tag{3.19}$$

If we now expand each of the cofactors (all determinants of order 2) according to equation (3.7), we obtain the full expansion of the determinant, given in equation (3.20), expressed as a sum of three positive and three negative terms:

$$\begin{vmatrix} a_{11} & a_{12} & a_{13} \\ a_{21} & a_{22} & a_{23} \\ a_{31} & a_{32} & a_{33} \end{vmatrix} = a_{11}a_{22}a_{33} - a_{11}a_{23}a_{32} - a_{12}a_{21}a_{33} + a_{12}a_{23}a_{31} + a_{13}a_{21}a_{32} - a_{13}a_{22}a_{31} \tag{3.20}$$

In this example, the determinant is initially expanded from the first row, but in fact we could just as easily expand from any row or column. Thus, for example, expanding from column two gives the alternative expansion:

$$\begin{vmatrix} a_{11} & a_{12} & a_{13} \\ a_{21} & a_{22} & a_{23} \\ a_{31} & a_{32} & a_{33} \end{vmatrix} = a_{12}A_{12} + a_{22}A_{22} + a_{32}A_{32} \tag{3.21}$$

which, upon expanding the cofactors, yields:

$$\begin{vmatrix} a_{11} & a_{12} & a_{13} \\ a_{21} & a_{22} & a_{23} \\ a_{31} & a_{32} & a_{33} \end{vmatrix} = -a_{12}\begin{vmatrix} a_{21} & a_{23} \\ a_{31} & a_{33} \end{vmatrix} + a_{22}\begin{vmatrix} a_{11} & a_{13} \\ a_{31} & a_{33} \end{vmatrix} - a_{32}\begin{vmatrix} a_{11} & a_{13} \\ a_{21} & a_{23} \end{vmatrix} \quad (3.22)$$

Expanding each of the determinants of order 2 in equation (3.22) yields equation (3.20), but with the six terms on the right in a different order.

A slightly quicker route to ensuring the correct signs in the sum of the cofactor values is obtained by remembering the general rule for expansion from any row or column in pictorial form:

$$\begin{vmatrix} + & - & + & - & + & \dots \\ - & + & - & + & - & \dots \\ + & - & + & - & \dots & \dots \\ - & + & - & \dots & \dots & \dots \\ + & - & \dots & \dots & \dots & \dots \end{vmatrix} \quad (3.23)$$

We shall discuss cofactors again when we meet matrix inverses in Chapter 4.

Worked Problem 3.1

Q Expand the following determinants from the given row or column, as indicated:

(a) $\begin{vmatrix} 0 & 1 \\ 1 & 0 \end{vmatrix}$; (b) $\begin{vmatrix} \cos\theta & -\sin\theta & 0 \\ \sin\theta & \cos\theta & 0 \\ 0 & 0 & 1 \end{vmatrix}$, from column 3.

A From the definition given for the expansion of a determinant of order 2, we have:

(a) $\begin{vmatrix} 0 & 1 \\ 1 & 0 \end{vmatrix} = 0 - 1 \times 1 = -1$

(b) $$\begin{vmatrix} \cos\theta & -\sin\theta & 0 \\ \sin\theta & \cos\theta & 0 \\ 0 & 0 & 1 \end{vmatrix} = 0 + 0 + 1 \times \begin{vmatrix} \cos\theta & -\sin\theta \\ \sin\theta & \cos\theta \end{vmatrix}$$

$$= \cos^2\theta + \sin^2\theta = 1.$$

Problem 3.3

Expand $\begin{vmatrix} 1 & -1 & 2 \\ 0 & 3 & 0 \\ 2 & -2 & -2 \end{vmatrix}$ from (a) column 2 and (b) row 2.

Problem 3.4

(a) Evaluate the cofactors A_{33}, A_{22}, A_{32} and A_{23} of $\begin{vmatrix} 1 & 0 & -2 \\ 2 & 8 & 4 \\ 3 & 2 & 2 \end{vmatrix}$.

(b) Expand the cofactors A_{12} and A_{21} of $\begin{vmatrix} \cos\theta & -\sin\theta & 0 \\ \sin\theta & \cos\theta & 0 \\ 0 & 0 & 1 \end{vmatrix}$.

3.3 Properties of Determinants

(1) A determinant is unaltered in value if *all* rows and columns are interchanged, *e.g.*:

$$\begin{vmatrix} 1 & 2 \\ 3 & 4 \end{vmatrix} = \begin{vmatrix} 1 & 3 \\ 2 & 4 \end{vmatrix} = -2 \qquad (3.24)$$

A determinant can only have a value if the elements are numbers.

(2) A determinant changes sign if two rows or columns are interchanged:

$$\begin{vmatrix} 1 & 2 \\ 3 & 4 \end{vmatrix} = -\begin{vmatrix} 2 & 1 \\ 4 & 3 \end{vmatrix} = -2 \qquad (3.25)$$

(3) A factor can be removed from each element of *one* row (or column) to give a new determinant, the value of which when multiplied by the factor gives the original value of the determinant. For example:

$$\begin{vmatrix} 1 & 2 \\ 3 & 4 \end{vmatrix} = 2\begin{vmatrix} 1 & 1 \\ 3 & 2 \end{vmatrix} = -2 \qquad (3.26)$$

Here, the factor 2 has been removed from column 2. Conversely, when a determinant is multiplied by a constant, the constant can be absorbed into the determinant by multiplying the elements of *one* row (or column) by that constant.

(4) The value of a determinant is unaltered if a constant multiple of one row or column is added to or subtracted from another row or column, respectively. For example, if we subtract twice column 1 from column 2, we obtain:

$$\begin{vmatrix} 1 & 2 \\ 3 & 4 \end{vmatrix} = \begin{vmatrix} 1 & 0 \\ 3 & -2 \end{vmatrix} = -2 \qquad (3.27)$$

3.4 Strategies for Expanding Determinants where $n > 3$

We can take a number of different approaches for evaluating determinants of higher order:

(a) For determinants of 3 or lower order, it is easiest to expand in full from the row or column containing the greatest number of zeros (for example, see Problem 3.3).

(b) For determinants of order 4 to about 6, it is best to introduce as many zeros as possible to the right (or left) of the leading diagonal using properties (1)–(4) (Section 3.3). If all the elements to the right or left of the leading diagonal are zero, then:

$$\begin{vmatrix} a_{11} & 0 & 0 & \cdots & 0 \\ a_{21} & a_{22} & 0 & \cdots & 0 \\ a_{31} & a_{32} & a_{33} & \cdots & 0 \\ \vdots & \vdots & \vdots & \ddots & \vdots \\ a_{n1} & a_{n2} & a_{n3} & \cdots & a_{nn} \end{vmatrix} = a_{11}a_{22}a_{33} \cdots a_{nn} \tag{3.28}$$

and the expansion of the determinant is given by the product of the elements lying on the leading diagonal.

(c) When expanding determinants of high order ($n > 5$), it is best to use one of the widely available computer algebra systems (Maple, Mathematica, *etc.*) or a numerical computer algorithm. There are many chemical situations in which we have to expand determinants of large order. For example, in computing the vibrational frequencies of ethene, it is necessary to expand a determinant of order 12 (for a non-linear molecule containing N nuclei, the order will be $3N - 6$).

Problem 3.5

Expand each of the following determinants:

(a) $\begin{vmatrix} 1 & 2 & 3 \\ 0 & 8 & 2 \\ -2 & 4 & 2 \end{vmatrix}$; (b) $\begin{vmatrix} 1 & 0 & -2 \\ 2 & 8 & 4 \\ 3 & 2 & 2 \end{vmatrix}$: (i) from row 2; (ii) using row/column operations to transform the first row to 1 0 0 before expansion from row 1; (iii) using row/column operations to transform all the elements to the right of the leading diagonal to zero, before expanding from row 1.

Problem 3.6

In using the Hückel model for calculating the molecular orbital energies, ε, for electrons in the π shell of the allyl system, it is necessary to solve the following equation:

$$\begin{vmatrix} \alpha-\varepsilon & \beta & 0 \\ \beta & \alpha-\varepsilon & \beta \\ 0 & \beta & \alpha-\varepsilon \end{vmatrix} = 0 \tag{3.29}$$

where the symbols α and β are parameters (both negative in value) of the model.

(a) Use property (3) of determinants to remove a factor of β from each row (or column) of the determinant shown in equation (3.29).

Hint: division of each element in *one* row or column by β results in a new determinant, the value of which is multiplied by β. Thus division of every element in the determinant results in a new determinant, the value of which is multiplied by β^3.

(b) Show that, on making the substitution $x = (\alpha - \varepsilon)/\beta$, the expansion of the determinant yields $x^3 - 2x = 0$.
(c) Find the three roots of this equation.
(d) Deduce the three orbital energies.

Summary of Key Points

This chapter develops the concept of the determinant as a precursor to a more complete treatment of matrix algebra in Chapter 4. The key points discussed include:

1. The use of determinants to solve sets of simultaneous linear equations.

2. The expansion of determinants of low order in full.

3. Cofactors of determinants.

4. Properties of determinants, and their use to simplify the expansion of determinants of high order.

4

Working with Arrays II: Matrices and Matrix Algebra

In the previous chapter we saw how determinants are used to tackle problems involving the solution of systems of linear equations. In general, the branch of mathematics which deals with linear systems is known as **linear algebra**, in which matrices and vectors play a dominant role. In this chapter we shall explore how matrices and matrix algebra are used to address problems involving coordinate transformations, as well as revisiting the solution of sets of simultaneous linear equations. Vectors are explored in Chapter 5.

Matrices are two-dimensional arrays (or tables) with specific shapes and properties:

$$\begin{pmatrix} 2 & -1 \\ 0 & 3 \end{pmatrix}, \quad \begin{pmatrix} x_1 \\ x_2 \\ x_3 \end{pmatrix}, \quad \begin{pmatrix} 1 & 2 & 3 \end{pmatrix}$$

Their key property is that they give us a formalism for systematically handling sets of objects – called **elements** – which, for example, can be numbers, chemical property values, algebraic quantities or integrals. Superficially, matrices resemble determinants, insofar as they are constructed from arrays of elements; however, as we shall see, they are really quite distinct from one another. The most important difference is that while a determinant expands to yield an expression (and a value, when its elements are numbers), a matrix does not!

Aims

In this chapter we develop matrix algebra from two key perspectives: one makes use of matrices to facilitate the handling of coordinate transformations, in preparation for a development of symmetry theory; the other revisits determinants and, through the definition of the matrix inverse, provides a means for solving sets of linear equations. By the end of this chapter, you should:

- Recognize the difference between a matrix and a determinant
- Recognize how matrices can be used to handle large linear systems in a compact way
- Be comfortable working with basic operations of matrix algebra (addition, subtraction, multiplication)
- Recognize specific kinds of matrix
- Use the special properties of a square matrix to evaluate its determinant and inverse
- Understand the basic principles of group theory

4.1 Introduction: Some Definitions

A **matrix** is an array of **elements**, comprising n rows and m columns, enclosed in parentheses (round brackets). By convention, matrices are named using bold typeface letters of upper or lower case, such as **A** or **b**, so we could, for example, label the matrices above as:

$$\mathbf{B} = \begin{pmatrix} 2 & -1 \\ 0 & 3 \end{pmatrix}, \quad \mathbf{c} = \begin{pmatrix} x_1 \\ x_2 \\ x_3 \end{pmatrix}, \quad \mathbf{d} = \begin{pmatrix} 1 & 2 & 3 \end{pmatrix}$$

The elements of the matrix are usually denoted a_{ij} or b_{ij} (depending on the letter used to label the matrix itself), where i denotes the row and j the column number. Thus, for example, the matrix **B** above has two rows and two columns, and is said to be a 2×2 matrix; however, as the matrix is square, it is sometimes named a **square matrix** of **order** n, with elements assigned as follows:

$$b_{11} = 2, \; b_{12} = -1, \; b_{21} = 0, \; b_{22} = 3$$

Sometimes, it is more convenient to use the notation $(\mathbf{B})_{ij}$ to indicate the ijth element of matrix **B**. Similarly, the 3×1 matrix **c** is called a **column matrix**, and the 1×3 matrix **d** is called a **row matrix**. The general matrix, **A**, having order $(n \times m)$, is called a **rectangular matrix** with elements:

$$\mathbf{A} = \begin{pmatrix} a_{11} & a_{12} & \cdots & a_{1m} \\ a_{21} & a_{22} & \cdots & a_{2m} \\ \vdots & \vdots & \vdots & \vdots \\ a_{n1} & a_{n2} & \cdots & a_{nm} \end{pmatrix} \tag{4.1}$$

Two matrices **A** and **B** are equal if, and only if, $a_{ij} = b_{ij}$ for all i,j. This also implies that the two matrices have the same order.

Problem 4.1

For each of the matrices $\mathbf{b} = \begin{pmatrix} 1 & 1 & 1 \\ 2 & -2 & 2 \end{pmatrix}$, $\mathbf{c} = \begin{pmatrix} 3 & -1 \\ 1 & -3 \end{pmatrix}$, $\mathbf{d} = \begin{pmatrix} 1 \\ 0 \end{pmatrix}$ and $\mathbf{e} = (0 \quad -\mathrm{i} \quad 1 \quad \mathrm{i})$: (a) name their shapes; (b) list their elements; (c) give their order.

4.2 Rules for Combining Matrices

In this section we explore the matrix analogues of addition, subtraction and multiplication of numbers. The analogue for division (the inverse operation of multiplication) has no direct counterpart for matrices.

4.2.1 Multiplication of a Matrix by a Constant

The multiplication of a matrix, **A**, by a constant c (a real, imaginary or complex number) is achieved by simply multiplying each element by the constant, resulting in the elements changing from a_{ij} to ca_{ij}, for all i,j.

Problem 4.2

Multiply the following matrices by 2:

(a) $\mathbf{B} = \begin{pmatrix} 4 & 5 \\ 1 & 6 \\ -4 & 3 \end{pmatrix}$; (b) $\mathbf{C} = \begin{pmatrix} 2 & \frac{5}{2} \\ \frac{1}{2} & 3 \\ -2 & \frac{3}{2} \end{pmatrix}$.

4.2.2 Addition and Subtraction of Matrices

If two matrices have the same order, then addition and subtraction are defined as:

$$\mathbf{C} = \mathbf{A} \pm \mathbf{B}, \text{ with } c_{ij} = a_{ij} \pm b_{ij}, \text{ for all } i,j \tag{4.2}$$

Neither addition nor subtraction is defined for combining matrices of different orders.

Worked Problem 4.1

Q Given the following matrices (notice that **A** has real, complex and imaginary elements):

$$\mathbf{A} = \begin{pmatrix} 2 & 2i & -3 \\ 1 & 2 & 8 \\ 5 & -4+i & 1 \end{pmatrix}, \mathbf{B} = \begin{pmatrix} 4 & 5 \\ 1 & 6 \\ -4 & 3 \end{pmatrix}, \mathbf{C} = \begin{pmatrix} 3 \\ 1 \\ 2 \end{pmatrix}, \mathbf{D} = \begin{pmatrix} 1 & 1 \\ 4 & 1 \\ 5 & 1 \end{pmatrix}$$

and $\mathbf{F} = \begin{pmatrix} 2 \\ 1 \\ 4 \end{pmatrix}$, evaluate:

(a) $\mathbf{G} = 2\mathbf{B} + \mathbf{D}$; (b) $\mathbf{M} = \mathbf{C} - 2\mathbf{F}$; (c) $\mathbf{H} = i\mathbf{A}$.

A (a) $\mathbf{G} = \begin{pmatrix} 8 & 10 \\ 2 & 12 \\ -8 & 6 \end{pmatrix} + \begin{pmatrix} 1 & 1 \\ 4 & 1 \\ 5 & 1 \end{pmatrix} = \begin{pmatrix} 9 & 11 \\ 6 & 13 \\ -3 & 7 \end{pmatrix}$.

(b) $\mathbf{M} = \begin{pmatrix} 3 \\ 1 \\ 2 \end{pmatrix} - 2\begin{pmatrix} 2 \\ 1 \\ 4 \end{pmatrix} = \begin{pmatrix} 3-4 \\ 1-2 \\ 2-8 \end{pmatrix} = \begin{pmatrix} -1 \\ -1 \\ -6 \end{pmatrix}$.

(c) $\mathbf{H} = i\begin{pmatrix} 2 & 2i & -3 \\ 1 & 2 & 8 \\ 5 & -4+i & 1 \end{pmatrix} = \begin{pmatrix} 2i & -2 & -3i \\ i & 2i & 8i \\ 5i & -4i-1 & i \end{pmatrix}$.

Problem 4.3

If $\mathbf{A} = \begin{pmatrix} 1 & i \\ -i & 1 \end{pmatrix}$, $\mathbf{B} = \begin{pmatrix} 1 & -i \\ i & 1 \end{pmatrix}$, $\mathbf{R} = \begin{pmatrix} \cos\theta & \sin\theta \\ -\sin\theta & \cos\theta \end{pmatrix}$ and $\mathbf{S} = \begin{pmatrix} \cos\theta & -\sin\theta \\ \sin\theta & \cos\theta \end{pmatrix}$, express:

(a) $\mathbf{A}+\mathbf{B}$, (b) $\mathbf{A}-\mathbf{B}$, (c) $\mathbf{R}+\mathbf{S}$ and (d) $\mathbf{R}-\mathbf{S}$ in terms of $\mathbf{C} = \begin{pmatrix} 0 & 1 \\ -1 & 0 \end{pmatrix}$ and $\mathbf{D} = \begin{pmatrix} 1 & 0 \\ 0 & 1 \end{pmatrix}$.

4.2.3 Matrix Multiplication

Given an $n \times m$ matrix, **A**, and an $m \times p$ matrix, **B**, then the *ij*th element, c_{ij}, of the resulting $n \times p$ product matrix $\mathbf{C} = \mathbf{AB}$, is found by selecting the i,j values and then, for each choice, summing the products of the elements in *row i* of **A** with those in *column j* of **B** (Figure 4.1):

$$\underset{(n\times p)}{\mathbf{C}} = \underset{(n\times m)}{\mathbf{A}} \quad \underset{(m\times p)}{\mathbf{B}} \tag{4.3}$$

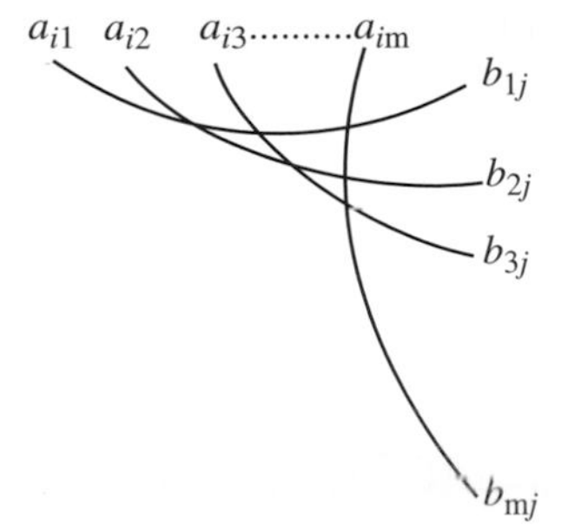

Figure 4.1 The product of an $n \times m$ matrix, **A**, and an $m \times p$ matrix, **B**, is an $n \times p$ matrix, **C**, whose *ij*th element, c_{ij}, is found by summing the products of the elements in *row i* of **A** with those in *column j* of **B**

A number of features relating to matrix multiplication are worthy of note:

- If the number of columns in **A** is not equal to the number of rows in **B**, then multiplication is undefined.
- In general, even if **AB** is defined, then **BA** may not be defined.
- If **AB** and **BA** are both defined, their orders may differ.
- Even if **AB** and **BA** have the same order, the two product matrices may not be equal. In these circumstances, matrix multiplication is non-commutative, *i.e.* AB ≠ BA.

Worked Problem 4.2

Q Where defined, determine the products (a) **AB** and (b) **BA** of the following matrices:

$$\mathbf{A} = \begin{pmatrix} 1 & 3 \\ 3 & 1 \end{pmatrix} \quad \text{and} \quad \mathbf{B} = \begin{pmatrix} 1 & 1 & 2 \\ 1 & 2 & 1 \end{pmatrix}$$

A (a) Matrix **A** is 2 × 2 and **B** is 2 × 3; thus the product **AB** is defined, as the number of columns in **A** is same as the number of rows in **B**. The product matrix will have order (2 × 3) (the number of rows in **A** and the number of columns in **B**):

$$\begin{aligned} \mathbf{AB} &= \begin{pmatrix} 1 & 3 \\ 3 & 1 \end{pmatrix}\begin{pmatrix} 1 & 1 & 2 \\ 1 & 2 & 1 \end{pmatrix} \\ &= \begin{pmatrix} (1\times1)+(3\times1) & (1\times1)+(3\times2) & (1\times2)+(3\times1) \\ (3\times1)+(1\times1) & (3\times1)+(1\times2) & (3\times2)+(1\times1) \end{pmatrix} \\ &= \begin{pmatrix} 4 & 7 & 5 \\ 4 & 5 & 7 \end{pmatrix}. \end{aligned}$$

(b) **BA** is undefined because the number of columns in **B** is not same as the number of rows in **A**.

Problem 4.4

For the matrices:
$\mathbf{A} = \begin{pmatrix} 1 & 2 \\ 2 & 1 \end{pmatrix}$, $\mathbf{B} = \begin{pmatrix} 1 & -1 \\ -1 & 2 \end{pmatrix}$, $\mathbf{C} = \begin{pmatrix} -1 & 1 \\ -1 & 1 \end{pmatrix}$, $\mathbf{D} = \begin{pmatrix} 1 & 2 \end{pmatrix}$ and $\mathbf{E} = \begin{pmatrix} 3 \\ -1 \end{pmatrix}$, find each product matrix specified below, where defined, and give its order, as appropriate:
AB, **BA**, **AC**, **BC**, **DE**, **ED**, **DA**, **AD**, **EA**, **AE**, **AB** – **BA**, (**AB**)**C**, **A**(**BC**), **A**(**B** + **C**), **AB** + **AC**.

Properties of Matrix Multiplication

You may have observed from your answers to Problem 4.4 that multiplication of matrices follows similar rules to that of numbers, insofar as it is:

- **Associative**:

$$(\mathbf{AB})\mathbf{C} = \mathbf{A}(\mathbf{BC}) \tag{4.4}$$

- **Distributive**:

$$\mathbf{A}(\mathbf{B} + \mathbf{C}) = \mathbf{AB} + \mathbf{AC} \tag{4.5}$$

One exception is the **commutative law**. In general, matrix multiplication is:
- Non-commutative:

$$\mathbf{AB} \neq \mathbf{BA} \quad \text{(except in certain special situations)} \tag{4.6}$$

As we suggested earlier, there is no general way of defining matrix division; however, for some square matrices we can define an operation that looks superficially like division, but it is really only multiplication (see Section 4.6).

4.3 Origins of Matrices

4.3.1 Coordinate Transformations

Matrices have their origin in **coordinate transformations**, where, in two dimensions, for example, a chosen point, with coordinates (x,y), is transformed to a new location with coordinates (x',y'). For example, consider an anticlockwise rotation of the point P in the xy-plane, about the z-axis, through an angle θ, as shown in Figure 4.2.

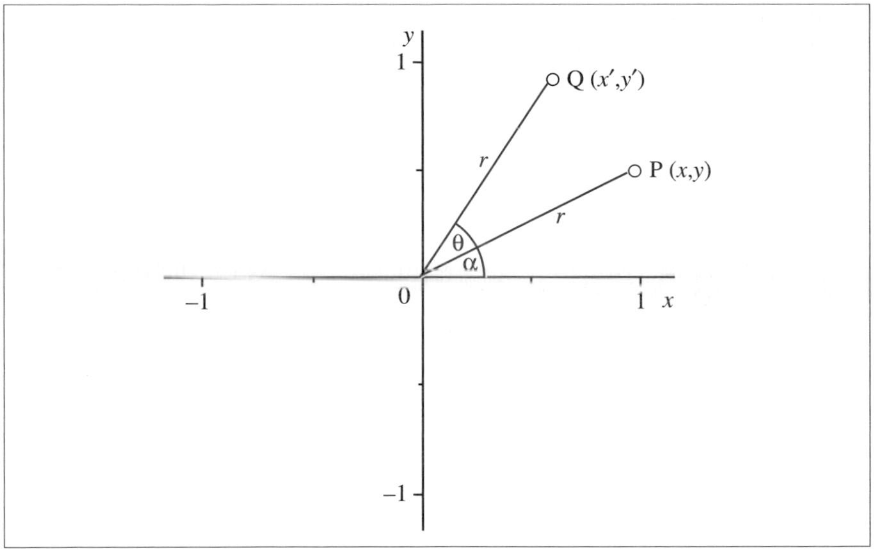

Figure 4.2 Rotation about the *z*-axis of the point P (*x*,*y*), through an angle θ to Q (x', y'); α is the angle between OP and the *x*-axis

We can use simple trigonometry to relate the coordinates of Q to those of P by expressing the Cartesian coordinates in terms of polar coordinates. Thus, the (x,y) coordinates of point P become:

$$x = r\cos\alpha \quad \text{and} \quad y = r\sin\alpha \tag{4.7}$$

and those of point Q become:

$$x' = r\cos(\theta+\alpha) \quad \text{and} \quad y' = r\sin(\theta+\alpha) \tag{4.8}$$

If we now use the addition theorems for cosine and sine (see Volume 1, Section 2.3.3), we obtain the expansions:

$$x' = r\cos(\theta+\alpha) = r\cos\theta\cos\alpha - r\sin\theta\sin\alpha = x\cos\theta - y\sin\theta \tag{4.9}$$

$$y' = r\sin(\theta+\alpha) = r\sin\theta\cos\alpha + r\cos\theta\sin\alpha = x\sin\theta + y\cos\theta \tag{4.10}$$

which allows us to express the coordinates of Q (x', y') in terms of those of P (x,y):

$$\begin{aligned} x' &= x\cos\theta - y\sin\theta \\ y' &= x\sin\theta + y\cos\theta \end{aligned} \tag{4.11}$$

Equation (4.11) describes the transformation of coordinates under an anticlockwise rotation by an angle, θ. This coordinate transformation is completely characterized by a square matrix, **A**, with the elements $\cos\theta$ and $\pm\sin\theta$, and the column matrices, **r** and **r**′, involving the initial and final coordinates, respectively:

$$\mathbf{A} = \begin{pmatrix} \cos\theta & -\sin\theta \\ \sin\theta & \cos\theta \end{pmatrix}, \ \mathbf{r}' = \begin{pmatrix} x' \\ y' \end{pmatrix}, \ \mathbf{r} = \begin{pmatrix} x \\ y \end{pmatrix} \tag{4.12}$$

We can now use matrix notation to replace the two equations (4.11) by the single matrix equation (4.13):

$$\begin{pmatrix} x' \\ y' \end{pmatrix} = \begin{pmatrix} \cos\theta & -\sin\theta \\ \sin\theta & \cos\theta \end{pmatrix} \begin{pmatrix} x \\ y \end{pmatrix} \qquad (4.13)$$
$$\mathbf{r}' \quad = \quad \mathbf{A} \quad \mathbf{r}$$

We can confirm that equation (4.13) correctly represents the coordinate transformation by evaluating the product matrix **Ar** on the right side:

$$\begin{pmatrix} x' \\ y' \end{pmatrix} = \begin{pmatrix} x\cos\theta - y\sin\theta \\ x\sin\theta + y\cos\theta \end{pmatrix} \qquad (4.14)$$
$$\mathbf{r}' \quad = \quad \mathbf{Ar}$$

Since $\mathbf{r}'$ and **Ar** are both 2×1 matrices, we can equate the elements in $\mathbf{r}'$ with those in **Ar**, to restore the original equations, which confirms equation (4.13) as the correct matrix representation of equations (4.11).

Worked Problem 4.3

Consider the coordinate transformation involving reflection in the y-axis (Figure 4.3). We can see that this transformation simply involves a change in sign of x, with the value of y remaining unchanged. Thus the transformed point, Q, will have coordinates $(x', y') = (-x, y)$.

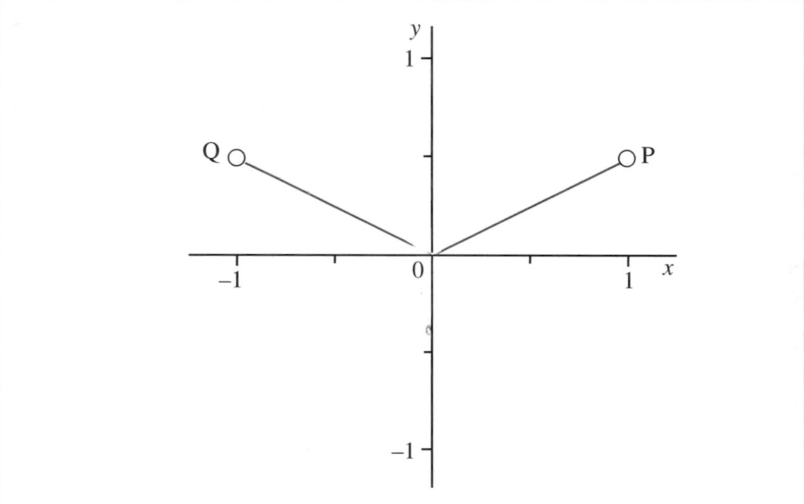

Figure 4.3 Reflection of the point P (x,y), in the y-axis to obtain the point Q $(-x,y)$

We can represent this transformation in terms of a matrix equation $\mathbf{r}' = \mathbf{Cr}$, where $\mathbf{C}$ is a 2×2 matrix characterizing reflection in the y-axis, and $\mathbf{r}$ and $\mathbf{r}'$ are the column matrices containing the initial and final coordinates, respectively.

Q Show that the matrix $\mathbf{C}$ characterizing reflection in the y-axis is $\mathbf{C} = \begin{pmatrix} -1 & 0 \\ 0 & 1 \end{pmatrix}$

A The coordinate transformation written as a matrix equation is:

$$\underset{\mathbf{r}'}{\begin{pmatrix} x' \\ y' \end{pmatrix}} = \begin{pmatrix} -x \\ y \end{pmatrix} = \underset{\mathbf{C}}{\begin{pmatrix} c_{11} & c_{12} \\ c_{21} & c_{22} \end{pmatrix}} \underset{\mathbf{r}}{\begin{pmatrix} x \\ y \end{pmatrix}}$$

where c_{11}, c_{12}, c_{21} and c_{22} are the elements of the matrix $\mathbf{C}$ which characterize reflection in the y-axis. Multiplying out the right side, we have:

$$c_{11}x + c_{12}y = -x \quad \text{and} \quad c_{21}x + c_{22}y = y$$

Thus, if we compare the x and y coefficients on each side of these equations, we obtain:

$$c_{11} = -1,\ c_{12} = 0,\ c_{21} = 0 \quad \text{and} \quad c_{22} = 1 \text{ and so } \mathbf{C} = \begin{pmatrix} -1 & 0 \\ 0 & 1 \end{pmatrix}$$

Sequential Coordinate Transformations

The effect of applying two sequential coordinate transformations on a point, $\mathbf{r}$, can be represented by the product of the two matrices, each one of which represents the respective transformation. We need to take care, however, that the matrices are multiplied in the correct order because, as we saw above, matrix multiplication is often non-commutative. For example, in order to find the matrix representing an anticlockwise rotation by θ, followed by a reflection in the y-axis, we need to find the product $\mathbf{CA}$ (and not $\mathbf{AC}$ as we might initially assume!).

Problem 4.5

(a) Find the matrix, $\mathbf{D}$, describing the coordinate transformation resulting from reflection in the line $y = x$.
(b) (i) Find the matrix, $\mathbf{E}$, describing the coordinate transformation resulting from a reflection in the line $y = x$, followed by a reflection in the y-axis (see Worked Problem 4.3). (ii) Find the matrix,

F, describing the coordinate transformation that results from a reflection in the y-axis followed by a reflection in the line $y = x$.

In each case, check your answer graphically, by using the matrices **E** and **F** to transform the coordinates (1, 2) to their new location (x', y').

4.3.2 Coordinate Transformations in Three Dimensions: A Chemical Example

In preparation for the discussion of group theory in Section 4.9, let us consider how we might use matrix representations of coordinate transformations to characterize the shape of a molecule – something of vital importance, for example, in describing the vibrational motions in a molecule. In order to accomplish this objective, we need to consider only those linear transformations in three dimensions that interchange equivalent points in a molecule. One example of such a transformation involves the interchange of coordinates defining the positions of two fluorine nuclei in the planar molecule BF_3. We can achieve this result by extending the rotation and reflection coordinate transformations in Figures 4.2 and 4.3 to three dimensions. For BF_3, there are four mirror planes, but, for the moment, let us focus only on the yz mirror plane, which is perpendicular to the plane of the molecule and contains the boron and the fluorine nucleus F_1 (Figure 4.4).

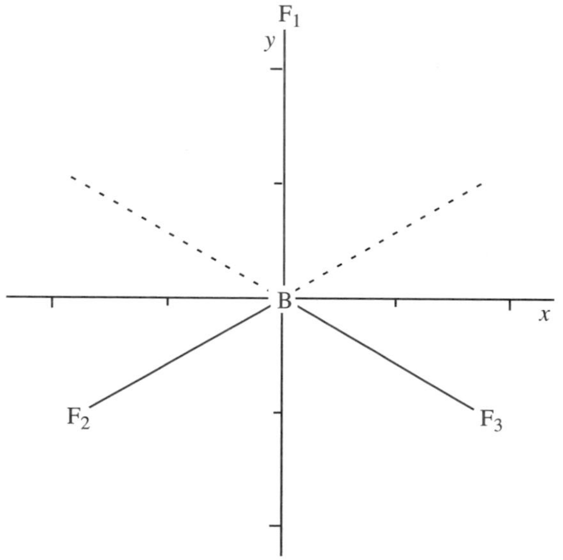

Figure 4.4 The nuclear configuration for BF_3 in the xy-plane. The z-axis is perpendicular to the paper, and passes through B. Three of the mirror planes are perpendicular to the paper, and contain the boron and one of the fluorine nuclei, respectively; the fourth mirror plane is lies in the plane of the molecule and contains all of the nuclei

The matrix **C,** defined in Worked Problem 4.3, describes reflection in the y-axis of a point defined by the two coordinates (x,y). We can rewrite matrix **C** in terms of all three coordinates as follows:

$$\begin{pmatrix} x' \\ y' \\ z' \end{pmatrix} = \begin{pmatrix} -1 & 0 & 0 \\ 0 & 1 & 0 \\ 0 & 0 & 1 \end{pmatrix} \begin{pmatrix} x \\ y \\ z \end{pmatrix} \qquad (4.15)$$
$$\mathbf{r}' = \mathbf{C} \quad \mathbf{r}$$

where we note that the z-coordinate is unchanged by the coordinate transformation. Thus, a reflection in the yz-plane interchanges points located at the nuclei F_2 and F_3.

If we now rotate an arbitrary point (x,y,z) about the z-axis, the x- and y-coordinates are transformed according to matrix **A**, defined in equation (4.12), but the z-coordinate is unchanged; thus:

$$\begin{pmatrix} x' \\ y' \\ z' \end{pmatrix} = \begin{pmatrix} \cos\theta & -\sin\theta & 0 \\ \sin\theta & \cos\theta & 0 \\ 0 & 0 & 1 \end{pmatrix} \begin{pmatrix} x \\ y \\ z \end{pmatrix} \qquad (4.16)$$
$$\mathbf{r}' = \mathbf{B} \quad \mathbf{r}$$

An anticlockwise rotation of $\theta = 2\pi/3$ (equivalent to 120°) about the z-axis described in equation (4.16) transforms a point located at either F_1, F_2 or F_3 to an equivalent point located at F_2, F_3 or F_1, respectively.

The important point here is that if the coordinates of points are represented in matrix form, then the geometrical actions involved in carrying out a rotation or reflection may also be represented by matrices, which enables us to mimic problems in geometry using matrix algebra; that is, geometrical operations on points can be replaced by **matrix representations** acting on column matrices containing the coordinates of points. We shall re-visit these ideas in Section 4.9, where we develop a brief introduction to the principles of symmetry theory.

4.4 Operations on Matrices

4.4.1 The Transpose of a Matrix

Given an $n \times m$ matrix, **B**, we can construct its **transpose**, $\mathbf{B}^{\mathrm{T}}$, by interchanging the rows and columns. Thus the ijth element of **B** becomes the jith element of $\mathbf{B}^{\mathrm{T}}$ according to:

$$(\mathbf{B})_{ij} = (\mathbf{B}^{\mathrm{T}})_{ji} \qquad (4.17)$$

Worked Problem 4.4

Q Find the transpose of: (a) $\mathbf{B} = \begin{pmatrix} 1 & 1 \\ 3 & 2 \\ 4 & 1 \end{pmatrix}$; (b) $\mathbf{C} = (0 \quad -1 \quad 1)$.

A $\mathbf{B}^T = \begin{pmatrix} 1 & 3 & 4 \\ 1 & 2 & 1 \end{pmatrix}$, (b) $\mathbf{C}^T = \begin{pmatrix} 0 \\ -1 \\ 1 \end{pmatrix}$.

Problem 4.6

Find the transpose of each of the following matrices:

(a) $\mathbf{A} = \begin{pmatrix} \cos\theta & -\sin\theta \\ \sin\theta & \cos\theta \end{pmatrix}$; (b) $\mathbf{C} = \begin{pmatrix} -1 & 1 \\ -1 & 1 \end{pmatrix}$; (c) $\mathbf{D} = \begin{pmatrix} 1 & 3 & 4 \\ 1 & 2 & 1 \end{pmatrix}$.

Problem 4.7

If $\mathbf{X}$ is an $n \times m$ matrix, then: (a) give the order of $\mathbf{XX}^\mathrm{T}$ and $\mathbf{X}^\mathrm{T}\mathbf{X}$; (b) use the matrix $\mathbf{B}$ from Worked Problem 4.2 to find $\mathbf{BB}^\mathrm{T}$ and $\mathbf{B}^\mathrm{T}\mathbf{B}$.

4.4.2 The Complex Conjugate Matrix

Taking the complex conjugate of every element of a matrix, $\mathbf{A}$, yields the **complex conjugate matrix**, $\mathbf{A}^*$; that is, $(\mathbf{A}^*)_{ij} = (\mathbf{A})_{ij}^*$. If all the elements of $\mathbf{A}$ are real, then $\mathbf{A}^* = \mathbf{A}$.

4.4.3 The Complex Conjugate Transposed Matrix

The transpose of the complex conjugate matrix (sometimes termed the **adjoint matrix**), is written as $\mathbf{A}^\dagger$ and defined such that:

$$\mathbf{A}^\dagger = (\mathbf{A}^*)^\mathrm{T} \equiv (\mathbf{A}^\mathrm{T})^* \Rightarrow (\mathbf{A}^\dagger)_{ij} = (\mathbf{A}^*)_{ji}. \tag{4.18}$$

If $\mathbf{A}^* = \mathbf{A}$ (a real matrix) then $\mathbf{A}^\dagger = \mathbf{A}^\mathrm{T}$.

Problem 4.8

If $\mathbf{A} = \begin{pmatrix} 1+\mathrm{i} & \mathrm{i} \\ -\mathrm{i} & 1 \end{pmatrix}$, give the forms of $\mathbf{A}^*$ and $\mathbf{A}^\dagger$.

Problem 4.9

Show that for the matrices $\mathbf{A} = \begin{pmatrix} 1 & 1 & \mathrm{i} \\ 1+\mathrm{i} & -1 \end{pmatrix}$ and $\mathbf{B} = \begin{pmatrix} 1 & 1+\mathrm{i} \\ 1+\mathrm{i} & 0 \end{pmatrix}$:

(a) $(\mathbf{AB})^* = \mathbf{A}^*\mathbf{B}^*$; (b) $(\mathbf{AB})^\dagger = \mathbf{B}^\dagger\mathbf{A}^\dagger$.

Note: these results are valid for any matrices **A** and **B**, for which multiplication is defined.

4.4.4 The Trace of a Square Matrix

The **trace** of a square matrix, **A**, of order n, denoted by tr**A**, is defined as the sum of its diagonal elements:

$$\mathrm{tr}\mathbf{A} = \sum_{i=1}^{n} (\mathbf{A})_{ii} \tag{4.19}$$

For example, the matrix $\mathbf{A} = \begin{pmatrix} 1 & -1 & 0 \\ 2 & -3 & 1 \\ 1 & -2 & 0 \end{pmatrix}$, has $\mathrm{tr}\mathbf{A} = 1 - 3 + 0 = -2$.

Since the transpose of a square matrix leaves the diagonal unchanged, we see that $\mathrm{tr}\mathbf{A} = \mathrm{tr}\mathbf{A}^\mathrm{T}$.

Problem 4.10

For the matrices:

$\mathbf{A} = \begin{pmatrix} 1 & -1 \\ 0 & 3 \end{pmatrix}$, $\mathbf{B} = \begin{pmatrix} 0 & 1 \\ 1 & -2 \end{pmatrix}$, $\mathbf{C} = \begin{pmatrix} -1 & 1 \\ 1 & 0 \end{pmatrix}$ and $\mathbf{D} = \begin{pmatrix} 1 & -1 & 0 \\ 1 & -2 & 0 \end{pmatrix}$,

show that:

(a) $\mathrm{tr}(\mathbf{AB}) = \mathrm{tr}(\mathbf{BA})$; (b) $\mathrm{tr}(\mathbf{ABC}) = \mathrm{tr}(\mathbf{CAB}) = \mathrm{tr}(\mathbf{BCA})$; (c) $\mathrm{tr}(\mathbf{D}^\mathrm{T}\mathbf{D}) = \mathrm{tr}(\mathbf{DD}^\mathrm{T})$.

4.4.5 The Matrix of Cofactors

The cofactor of a determinant, which we first defined in Section 3.2, is characterized by a row and column index, in much the same way as we characterize the elements in a matrix. Thus, we can form the matrix of cofactors by accommodating each cofactor in its appropriate position. For example, the determinant:

$$\det \mathbf{A} = \begin{vmatrix} a_{11} & a_{12} \\ a_{21} & a_{22} \end{vmatrix} \tag{4.20}$$

gives rise to the four cofactors A_{11}, A_{12}, A_{21} and A_{22}, which may be collected together in the matrix of cofactors, **B**:

$$\mathbf{B} = \begin{pmatrix} A_{11} & A_{12} \\ A_{21} & A_{22} \end{pmatrix} = \begin{pmatrix} a_{22} & -a_{21} \\ -a_{12} & a_{11} \end{pmatrix} \tag{4.21}$$

4.5 The Determinant of a Product of Square Matrices

For two square matrices, **A** and **B**, of order n, the determinant of the product matrix **AB** is given by the product of the two determinants:

$$\det(\mathbf{AB}) = \det(\mathbf{BA}) = \det \mathbf{A} \times \det \mathbf{B} \tag{4.22}$$

We now return to a further discussion of some special matrices that arise in a chemical context.

4.6 Special Matrices

So far, we have met matrices of different orders, but we have not been concerned with the properties of their constituent elements. In this section, we introduce the **null** and **unit** matrices, and then present a catalogue of important kinds of matrix that are common in developing mathematical models used, for example, in the calculation of vibrational frequencies of molecules, distributions of electron density and other observable properties of molecules.

4.6.1 The Null Matrix

The general **null matrix** is an n by m matrix, all of whose elements are zero. If the matrix is:

- Rectangular, it is named as $\mathbf{O}_{nm}$.
- Square ($n = m$) it is named as $\mathbf{O}_n$.
- A column matrix, it is named $\mathbf{O}_{n1}$, or more commonly as **0**.

Given an $m \times n$ matrix **X**:

$$\mathbf{O}_{nm}\mathbf{X}_{mn} = \mathbf{O}_n, \qquad \mathbf{X}_{mn}\mathbf{O}_{nm} = \mathbf{O}_m \tag{4.23}$$

4.6.2 The Unit Matrix

The **unit matrix** is a square matrix of order n, denoted here by $\mathbf{E}_n$, whose leading diagonal elements are all unity (*i.e.* have value 1) and whose off-diagonal elements are zero. Thus, for example:

$$\mathbf{E}_2 = \begin{pmatrix} 1 & 0 \\ 0 & 1 \end{pmatrix}, \qquad \mathbf{E}_3 = \begin{pmatrix} 1 & 0 & 0 \\ 0 & 1 & 0 \\ 0 & 0 & 1 \end{pmatrix} \tag{4.24}$$

The elements of $\mathbf{E}_n$ may be denoted e_{ij}, but in practice they are usually specified using the **Kronecker delta**, which is written as:

$$e_{ij} = \delta_{ij} = \begin{cases} 1, & \text{for} \quad i=j \\ 0, & \text{for} \quad i \neq j \end{cases} \tag{4.25}$$

where $i=j$ designates a diagonal position and $i \neq j$ a non-diagonal position.

As $\mathbf{E}_n$ is an $n \times n$ matrix, $\mathbf{E}_n\mathbf{A}$ (pre-multiplication of $\mathbf{A}$ by $\mathbf{E}_n$) is equal to $\mathbf{A}$ if $\mathbf{A}$ has order $(n \times p)$; likewise, $\mathbf{A}\mathbf{E}_n$ (post-multiplication of $\mathbf{A}$ by $\mathbf{E}_n$) yields $\mathbf{A}$ if $\mathbf{A}$ has order $(p \times n)$.

Problem 4.11

For each of the following matrix products:

(a) $\mathbf{O}_{23}\begin{pmatrix} 1 & 3 \\ 2 & 2 \\ 0 & 1 \end{pmatrix}$; (b) $\begin{pmatrix} 1 & 2 & 3 \\ 4 & 5 & 6 \end{pmatrix}\mathbf{O}_{23}$; (c) $\mathbf{E}_3\begin{pmatrix} 1 & 3 \\ 2 & 2 \\ 0 & 1 \end{pmatrix}$; (d) $\begin{pmatrix} 1 & 2 & 3 \\ 4 & 5 & 6 \end{pmatrix}\mathbf{E}_3$;

(e) $\begin{pmatrix} 1 & 3 \\ 2 & 2 \\ 0 & 1 \end{pmatrix}\mathbf{E}_3$, give the resultant matrix, where the product is defined.

Problem 4.12

If $\mathbf{A} = \begin{pmatrix} \cos\theta & -\sin\theta \\ \sin\theta & \cos\theta \end{pmatrix}$ and $\mathbf{G} = \begin{pmatrix} 0 & -1 & 1 \\ 2 & -1 & 2 \\ 1 & 1 & 1 \end{pmatrix}$:

(a) (i) find det$\mathbf{A}$, and the matrix of cofactors, $\mathbf{B}$; (ii) show that $\mathbf{B}^{\mathrm{T}}\mathbf{A} = \mathbf{E}_2 \det\mathbf{A}$; (b) show that $\mathbf{H}^{\mathrm{T}}\mathbf{G} = \mathbf{E}_3 \det \mathbf{G}$, where $\mathbf{H}$ is the matrix of cofactors of det $\mathbf{G}$.

4.7 Matrices with Special Properties

4.7.1 Symmetric Matrices

A square matrix, **A**, is said to be **symmetric** if it has the property $(\mathbf{A})_{ij} = (\mathbf{A})_{ji}$; that is:

$$\mathbf{A} = \mathbf{A}^{\mathrm{T}} \tag{4.26}$$

For example, the following matrix is symmetric:

$$\mathbf{A} = \begin{pmatrix} 2 & 1 & 3 \\ 1 & 4 & 3 \\ 3 & 3 & 0 \end{pmatrix}$$

as reflection in the leading diagonal leaves the array of elements unchanged in appearance.

For any n by m matrix **X**, both $\mathbf{X}^{\mathrm{T}}\mathbf{X}$ and $\mathbf{X}\mathbf{X}^{\mathrm{T}}$ are symmetric matrices.

4.7.2 Orthogonal Matrices

An **orthogonal** matrix, **A**, is a square matrix of order n with the property:

$$\mathbf{A}^{\mathrm{T}}\mathbf{A} = \mathbf{A}\mathbf{A}^{\mathrm{T}} = \mathbf{E}_n \tag{4.27}$$

It follows from Property (1) of determinants (see Section 3.3), that $\det \mathbf{A} = \det \mathbf{A}^{\mathrm{T}}$, since the value of a determinant is unchanged if all columns and rows are interchanged. It also follows from equation (4.22) that $\det(\mathbf{A}\mathbf{A}^{\mathrm{T}}) = \det\mathbf{A} \times \det\mathbf{A} = (\det\mathbf{A})^2$ and from the property of an orthogonal matrix given in equation (4.27) that $(\det\mathbf{A})^2 = \det\mathbf{E}_n = 1$. Consequently, since $(\det\mathbf{A})^2 = 1$, it follows that, for an orthogonal matrix, $\det\mathbf{A} = \pm 1$. However, it does not necessarily follow that an arbitrary matrix satisfying this criterion is orthogonal, since it must also satisfy equation (4.27).

Problem 4.13

(a) Find the value(s) of k for which the matrix $\mathbf{A} = \begin{pmatrix} \frac{1}{\sqrt{2}} & k \\ \frac{1}{\sqrt{2}} & \frac{-1}{\sqrt{2}} \end{pmatrix}$ is orthogonal. Check your answer by verifying that $\mathbf{A}^{\mathrm{T}}\mathbf{A} = \mathbf{A}\mathbf{A}^{\mathrm{T}} = \mathbf{E}_n$.

(b) Find the value(s) of θ for which $\mathbf{R} = \begin{pmatrix} \cos\theta & \sin\theta & 0 \\ \sin\theta & \cos\theta & 0 \\ 0 & 0 & 1 \end{pmatrix}$ is orthogonal.
Hint: $\cos 2\theta = \cos^2\theta - \sin^2\theta$

(c) Find the value(s) of k for which the matrix $\mathbf{A} = \begin{pmatrix} 1 & k \\ -1 & 1 \end{pmatrix}$ satisfies the condition det $\mathbf{A} = \pm 1$. Check your answer against equation (4.27) and comment on whether the matrix $\mathbf{A}$ is orthogonal or not.

As we shall see in later sections, orthogonal matrices play an important role in defining the coordinate transformations that are used in characterizing the symmetry properties of molecules.

4.7.3 Singular Matrices

A square matrix, $\mathbf{A}$, for which det$\mathbf{A} = 0$, is said to be **singular**. Such matrices usually arise when the number of variables (or degrees of freedom) is over-specified for the chosen model, as would occur, for example, in:

- Using the same atomic orbital twice in constructing molecular orbitals in the Linear Combination of Atomic Orbitals (LCAO) model.
- Solving an inconsistent set of equations.

Superficially, using the same atomic orbital twice in constructing molecular orbitals using the LCAO method may seem misguided; however, there are cases when the second occurrence of the atomic orbital is disguised.

4.7.4 Hermitian Matrices

A complex square matrix, that is equal to the transpose of its complex conjugate, is called an **Hermitian** matrix; that is:

$$\mathbf{A} = \mathbf{A}^{\dagger} \tag{4.28}$$

Problem 4.14

(a) Verify that the matrix $\mathbf{A} = \begin{pmatrix} 0 & 3+\mathrm{i} \\ 3-\mathrm{i} & 1 \end{pmatrix}$ is Hermitian.

(b) If $\mathbf{x}$ is the 2 by 1 column matrix $\begin{pmatrix} 1 \\ \mathrm{i} \end{pmatrix}$, and $\mathbf{A}$ is the Hermitian matrix in part (a), show that $\mathbf{x}^{\dagger}\mathbf{A}\mathbf{x} = -1$.

4.7.5 Unitary Matrices

A square matrix $\mathbf{U}$, of order n, is said to be **unitary** if:

$$\mathbf{U}^{\dagger}\mathbf{U} = \mathbf{U}\mathbf{U}^{\dagger} = \mathbf{E}_n \tag{4.29}$$

It follows from the definition of a unitary matrix that $\det\mathbf{U} = \pm 1$. As for orthogonal matrices, an arbitrary matrix having the property $\det\mathbf{U} = \pm 1$ is unitary only if the requirements of equation (4.29) are satisfied.

Hermitian and unitary matrices play the same role for matrices with complex elements that symmetric and orthogonal matrices do for matrices with real elements. These features are summarized in Table 4.1.

Table 4.1 A summary of special square matrices

Matrices with real elements	*Matrices with complex elements*
Transpose, $\mathbf{A}^{\mathrm{T}}$ $\begin{pmatrix} 2 & 5 \\ 3 & 4 \end{pmatrix} \Rightarrow \begin{pmatrix} 2 & 3 \\ 5 & 4 \end{pmatrix}$	Complex conjugate transpose, $\mathbf{A}^{\dagger}$ $\begin{pmatrix} 2 & 3+\mathrm{i} \\ -\mathrm{i} & 4 \end{pmatrix} \Rightarrow \begin{pmatrix} 2 & \mathrm{i} \\ 3-\mathrm{i} & 4 \end{pmatrix}$
Symmetric $\mathbf{A} = \mathbf{A}^{\mathrm{T}}$ $\begin{pmatrix} 2 & 5 \\ 5 & 3 \end{pmatrix}$	Hermitian $\mathbf{A} = \mathbf{A}^{\dagger}$ $\begin{pmatrix} 2 & 3+\mathrm{i} \\ 3-\mathrm{i} & 1 \end{pmatrix}$
Orthogonal $\mathbf{A}^{\mathrm{T}}\mathbf{A} = \mathbf{A}\mathbf{A}^{\mathrm{T}} = \mathbf{E}_n$ $\begin{pmatrix} \cos\theta & -\sin\theta \\ \sin\theta & \cos\theta \end{pmatrix} \begin{pmatrix} \cos\theta & \sin\theta \\ -\sin\theta & \cos\theta \end{pmatrix} = \mathbf{E}_2$	Unitary $\mathbf{U}^{\dagger}\mathbf{U} = \mathbf{U}\mathbf{U}^{\dagger} = \mathbf{E}_n$ $\begin{pmatrix} \frac{1}{\sqrt{2}} & \frac{-\mathrm{i}}{\sqrt{2}} \\ \frac{\mathrm{i}}{\sqrt{2}} & \frac{-1}{\sqrt{2}} \end{pmatrix} \begin{pmatrix} \frac{1}{\sqrt{2}} & \frac{-\mathrm{i}}{\sqrt{2}} \\ \frac{\mathrm{i}}{\sqrt{2}} & \frac{-1}{\sqrt{2}} \end{pmatrix} = \mathbf{E}_2$
A consequence of the above is that $\det\mathbf{A} = \pm 1$	A consequence of the above is that $\det\mathbf{U} = \pm 1$

Symmetric, Hermitian, orthogonal and unitary matrices all arise in the quantum mechanical models used to probe aspects of molecular structure.

Problem 4.15

Classify each of the following matrices according to whether they are symmetric, Hermitian, orthogonal or unitary:

(a) $\mathbf{A} = \begin{pmatrix} 1 & \mathrm{i} \\ -\mathrm{i} & 0 \end{pmatrix}$; (b) $\mathbf{B} = \frac{1}{\sqrt{2}}\begin{pmatrix} 1 & \mathrm{i} \\ -\mathrm{i} & -1 \end{pmatrix}$; (c) $\mathbf{C} = \begin{pmatrix} 0 & -1 \\ 1 & 0 \end{pmatrix}$;

(d) $\mathbf{D} = \begin{pmatrix} 1 & -1 \\ -1 & 0 \end{pmatrix}$.

We now proceed to identify the last of the special matrices that are important to us.

4.7.6 The Inverse Matrix

The **inverse** of a square matrix **A**, of order *n*, is written $\mathbf{A}^{-1}$ and has the property:

$$\mathbf{A}\mathbf{A}^{-1} = \mathbf{A}^{-1}\mathbf{A} = \mathbf{E}_n \tag{4.30}$$

and exists only if det$\mathbf{A} \neq 0$. If det$\mathbf{A} = 0$, then **A** is singular and $\mathbf{A}^{-1}$ does not exist. We saw in Problem 4.12(a) that the transposed matrix of cofactors, $\mathbf{B}^{\mathrm{T}}$, is related to **A** and det**A** – irrespective of the order of **A** – according to the formula:

$$\mathbf{B}^{\mathrm{T}}\mathbf{A} = \mathbf{E}_n \det \mathbf{A} \tag{4.31}$$

Rearranging equation (4.31) gives:

$$\frac{1}{\det \mathbf{A}}\mathbf{B}^{\mathrm{T}} = \frac{\mathbf{E}_n}{\mathbf{A}} \tag{4.32}$$

but we know from equation (4.30) that $\mathbf{A}^{-1} = \dfrac{\mathbf{E}_n}{\mathbf{A}}$, and so:

$$\mathbf{A}^{-1} = \frac{1}{\det \mathbf{A}}\mathbf{B}^{\mathrm{T}} \tag{4.33}$$

which provides us with a formula for obtaining $\mathbf{A}^{-1}$ from $\mathbf{B}^{\mathrm{T}}$ and det **A**. It should be remembered, however, that $\mathbf{A}^{-1}$ exists only if **A** is non-singular.

Worked Problem 4.5

Q Find the inverse of the matrix $\mathbf{A} = \begin{pmatrix} 1 & -1 \\ 1 & 1 \end{pmatrix}$

A First, det $\mathbf{A} = 1 + 1 = 2$. The matrix of cofactors of **A** is:

$$\mathbf{B} = \begin{pmatrix} A_{11} & A_{12} \\ A_{21} & A_{22} \end{pmatrix} = \begin{pmatrix} 1 & -1 \\ 1 & 1 \end{pmatrix} \Rightarrow \mathbf{B}^{\mathrm{T}} = \begin{pmatrix} 1 & 1 \\ -1 & 1 \end{pmatrix}$$

Thus the inverse of matrix **A** is given by:

$$\mathbf{A}^{-1} = \frac{1}{2}\begin{pmatrix} 1 & 1 \\ -1 & 1 \end{pmatrix} = \begin{pmatrix} \frac{1}{2} & \frac{1}{2} \\ -\frac{1}{2} & \frac{1}{2} \end{pmatrix}$$

Check: use the definition of the matrix inverse to confirm that $\mathbf{A}\mathbf{A}^{-1} = \mathbf{A}^{-1}\mathbf{A} = \mathbf{E}_n$:

$$\begin{pmatrix} 1 & -1 \\ 1 & 1 \end{pmatrix}\begin{pmatrix} \frac{1}{2} & \frac{1}{2} \\ -\frac{1}{2} & \frac{1}{2} \end{pmatrix} = \begin{pmatrix} 1 & 0 \\ 0 & 1 \end{pmatrix} = \mathbf{E}_2$$

Problem 4.16

Find the inverse of the matrix $\mathbf{A} = \begin{pmatrix} 1 & -1 & 1 \\ -1 & -1 & 1 \\ 1 & 1 & 1 \end{pmatrix}$, and demonstrate that $\mathbf{AA}^{-1} = \mathbf{A}^{-1}\mathbf{A} = \mathbf{E}_3$.

The inverse matrix has many uses, but of particular relevance to us as chemists is the role they play in:

- The solution of sets of simultaneous linear equations.
- Developing the concept of a group which, in turn, underpins the basis of symmetry theory.

4.8 Solving Sets of Linear Equations

Suppose we have a set of three equations, each of which is linear in the unknowns x_1, x_2, x_3:

$$\begin{aligned} a_{11}x_1 + a_{12}x_2 + a_{13}x_3 &= b_1 \\ a_{21}x_1 + a_{22}x_1 + a_{23}x_1 &= b_2 \\ a_{31}x_1 + a_{32}x_1 + a_{33}x_1 &= b_3 \end{aligned} \tag{4.34}$$

where the a_{ij} and b_i ($i,j = 1, 2, 3$) are constant coefficients. If all the b_i are zero, then the equations are called **homogeneous**, but if one or more of the b_i are non-zero, then the equations are called **inhomogeneous**.

We can write the three linear equations (4.34) as a single matrix equation:

$$\begin{pmatrix} a_{11} & a_{12} & a_{13} \\ a_{21} & a_{22} & a_{23} \\ a_{31} & a_{32} & a_{33} \end{pmatrix} \begin{pmatrix} x_1 \\ x_2 \\ x_3 \end{pmatrix} = \begin{pmatrix} b_1 \\ b_2 \\ b_3 \end{pmatrix} \tag{4.35}$$

and then check that equations (4.35) and (4.34) are equivalent, by evaluating the matrix product in the left side of equation (4.35) to give:

$$\begin{pmatrix} a_{11}x_1 + a_{12}x_2 + a_{13}x_3 \\ a_{21}x_1 + a_{22}x_2 + a_{23}x_3 \\ a_{31}x_1 + a_{32}x_2 + a_{33}x_3 \end{pmatrix} = \begin{pmatrix} b_1 \\ b_2 \\ b_3 \end{pmatrix} \tag{4.36}$$

We now have two 3×1 matrices, which are equal to one another; because this implies equality of the elements, we regenerate the original linear equations given in equation (4.34). If we now rewrite equation (4.35) in a more compact form as:

$$\mathbf{Ax} = \mathbf{b} \tag{4.37}$$

and pre-multiply by $\mathbf{A}^{-1}$, the matrix inverse of $\mathbf{A}$, we obtain:

$$\mathbf{A}^{-1}\mathbf{A}\mathbf{x} = \mathbf{A}^{-1}\mathbf{b} \tag{4.38}$$

Since $\mathbf{A}^{-1}\mathbf{A} = \mathbf{E}_n$ and $\mathbf{E}_n\mathbf{x} = \mathbf{x}$, it follows that the unique solution is given by:

$$\mathbf{x} = \mathbf{A}^{-1}\mathbf{b} \tag{4.39}$$

However, this solution is meaningful only if det $\mathbf{A}$ is non-singular. If $\mathbf{A}$ is singular, the equations are inconsistent – in which case, no solution is forthcoming.

Worked Problem 4.6

Q (a) Confirm that the following equations have a single, unique solution:

$$\begin{aligned} x_1 - x_2 + x_3 &= 1 \\ -x_1 - x_2 + x_3 &= 2 \\ x_1 + x_2 + x_3 &= -1 \end{aligned} \tag{4.40}$$

(b) Find the solution.

A (a) Rewriting equation (4.40) in matrix form gives:

$$\underset{\mathbf{A}}{\begin{pmatrix} 1 & -1 & 1 \\ -1 & -1 & 1 \\ 1 & 1 & 1 \end{pmatrix}} \underset{\mathbf{x}}{\begin{pmatrix} x_1 \\ x_2 \\ x_3 \end{pmatrix}} = \underset{\mathbf{b}}{\begin{pmatrix} 1 \\ 2 \\ -1 \end{pmatrix}}$$

The set of equations has a unique solution as det $\mathbf{A} = 4$ (see Problem 4.16), indicating that the equations are consistent.
(b) Following the procedure in Worked Problem 4.5, and with reference to the answer to Problem 4.16, we find $\mathbf{A}^{-1} = \begin{pmatrix} \frac{1}{2} & -\frac{1}{2} & 0 \\ -\frac{1}{2} & 0 & \frac{1}{2} \\ 0 & \frac{1}{2} & \frac{1}{2} \end{pmatrix}$. Thus, the solution, according to equation (4.39), is given by:

$$\begin{pmatrix} x_1 \\ x_2 \\ x_3 \end{pmatrix} = \begin{pmatrix} \frac{1}{2} & -\frac{1}{2} & 0 \\ -\frac{1}{2} & 0 & \frac{1}{2} \\ 0 & \frac{1}{2} & \frac{1}{2} \end{pmatrix} \begin{pmatrix} 1 \\ 2 \\ -1 \end{pmatrix} = \begin{pmatrix} -\frac{1}{2} \\ -1 \\ \frac{1}{2} \end{pmatrix}$$

from which we see that $x_1 = -\frac{1}{2}$, $x_2 = -1$, $x_3 = \frac{1}{2}$.

Problem 4.17

Find the values of x, y, z that satisfy the equations:

$$\begin{aligned} x + 2y + 3z &= 1 \\ 8y + 2z &= 1 \\ -2x + 4y + 2z &= 2 \end{aligned}$$

So far we have considered only the solutions to sets of inhomogeneous linear equations where at least one of the b_i is non-zero. If, however, we have a set of homogeneous equations, where all the b_i are zero, then we may define two further possible limiting cases:

- If $\det \mathbf{A} \neq 0$ and $\mathbf{b} = \mathbf{0}$ (all b_i are zero), then this approach will only ever yield the solution $\mathbf{x} = \mathbf{0}$, *i.e.* $x_1 = x_2 = x_3 = 0$, since $\mathbf{x} = \mathbf{A}^{-1}\mathbf{0} = \mathbf{0}$.
- If $\det \mathbf{A} = 0$, and $\mathbf{b} = \mathbf{0}$, then, again, $\mathbf{A}^{-1}$ will be undefined. However, although the solution may yield the so-called trivial result $\mathbf{x} = \mathbf{0}$, other solutions may also exist.

4.8.1 Solution of Linear Equations: A Chemical Example

In Problem 3.6, we saw how the molecular orbital energies for the allyl system are determined from the solution of a determinantal equation. At this point, we are now in the position to understand the origin of this equation.

In the Hückel model, the result of minimizing the energy of the appropriately occupied π molecular orbitals results in the following set of linear equations in the unknown atomic orbital coefficients, c_r, together with the molecular orbital energy, ε:

$$\begin{aligned} c_1(\alpha - \varepsilon) + c_2\beta &= 0 \\ c_1\beta + c_2(\alpha - \varepsilon) + c_3\beta &= 0 \\ c_2\beta + c_3(\alpha - \varepsilon) &= 0 \end{aligned} \tag{4.41}$$

Equations (4.41) may be more succinctly expressed as a single matrix equation:

$$\begin{pmatrix} (\alpha - \varepsilon) & \beta & 0 \\ \beta & (\alpha - \varepsilon) & \beta \\ 0 & \beta & (\alpha - \varepsilon) \end{pmatrix} \begin{pmatrix} c_1 \\ c_2 \\ c_3 \end{pmatrix} = \mathbf{0} \tag{4.42}$$

or in more compact form as $\mathbf{Ac} = \mathbf{0}$, where $\mathbf{0}$ is a null column matrix. Equations (4.41) provide an example of a set of homogeneous equations, because the right-side constant coefficients, equivalent to b_i in equation (4.34), are all zero (and hence the appearance of the null column matrix $\mathbf{0}$

in the equivalent matrix equation 4.42). The trivial solution to equation (4.42), where $c_1 = c_2 = c_3 = 0$ ($\mathbf{c} = \mathbf{0}$) is of no physical significance, as the molecular orbitals do not then exist – another example of how important it is to use physical intuition to interpret the significance of a mathematical result! A more detailed study of the mathematics indicates that equation (4.42) has a non-trivial solution if det $\mathbf{A} = 0$, the solution of which yields the orbital energies $\varepsilon = \alpha, \mp \sqrt{2}\beta$, as seen in Problem 3.6.

Problem 4.10

The three molecular orbitals for the allyl system are obtained by solving the set of simultaneous equations (4.41) for each value of ε, in turn, to obtain the atomic orbital coefficients, c_i.

(a) For $\varepsilon = \alpha$, show that $c_3 = -c_1$ and that $c_2 = 0$.

(b) For $\varepsilon = \alpha + \sqrt{2}\beta$, show that $c_2 = \sqrt{2}c_1$ and $c_3 = c_1$.

(c) For $\varepsilon = \alpha - \sqrt{2}\beta$, show that $c_2 = -\sqrt{2}c_1$ and $c_3 = c_1$.

(d) For each of the three orbital energies, construct the column matrix, $\mathbf{c_i}$, where each element is expressed in terms of c_1.

4.9 Molecular Symmetry and Group Theory

One of the key applications of matrices in chemistry is in the characterization of molecular symmetry. In Section 4.3 we saw how it was possible to represent the coordinate transformations associated with rotation and reflection, in terms of matrices. These notions are now explored in the next section, where we develop some of the basic ideas of **group theory**.

4.9.1 An Introduction to Group Theory

A **group** consists of a **set** of **elements** (*e.g.* numbers or square matrices), for which there is a specified mode of combination (for example, addition, subtraction, multiplication), subject to the four following requirements:

(a) For any **R**, **S** in the set, the combination **RS** is a member of the set (**closure**).

(b) For any **R**, **S** and **T** in the set, the **mode of combination** must be associative, *i.e.* **R**(**ST**) = (**RS**)**T**.

(c) There is an **identity** element **E** such that, for any element, **R**, in the set, **RE** = **ER** = **R**.

7. The matrix of cofactors and the definition of the inverse matrix.

8. The application of matrix algebra for solving sets of simultaneous linear equations – homogeneous and inhomogeneous equations.

9. An introduction to molecular symmetry and group theory.

10. Elucidating the characteristic electronic structures associated with molecules.

11. Introducing some of the concepts necessary in the study and use of vectors.

5
Vectors

Many of the physical quantities which we deal with from day to day, such as mass, temperature or concentration, require only a single number (with appropriate units) to specify their value. Such quantities are called **scalar** quantities, specified exclusively in terms of their value. However, we frequently encounter other quantities, called **vectors**, which require us to specify a **magnitude** (a positive value) and a direction. Velocity is an example of a vector quantity, whereas speed is a scalar quantity (in fact speed defines the magnitude of velocity!). This is why we say that an object travelling on a circular path with constant speed (such as an electron orbiting a nucleus in the Bohr model of the atom) is accelerating: its velocity changes with time because its direction is constantly changing, in spite of the fact that the speed is constant (see Figure 5.1).

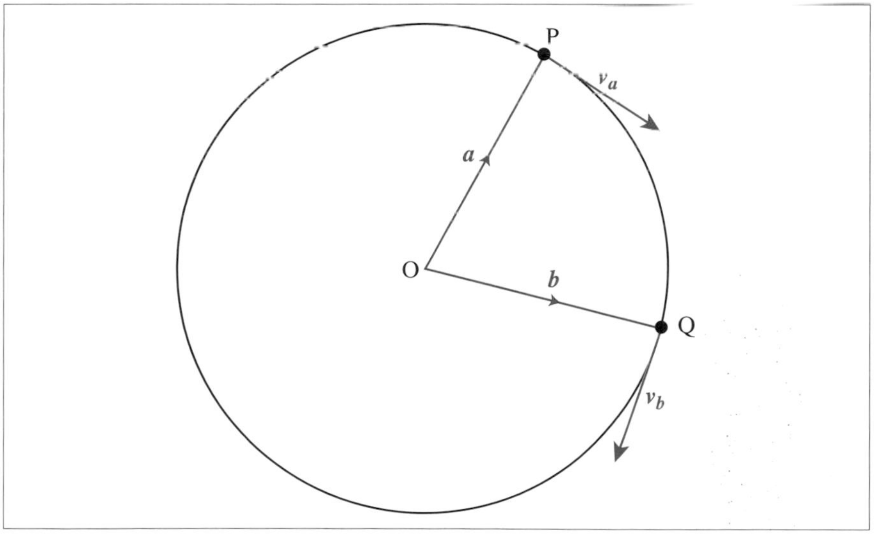

Figure 5.1 The velocity and position vectors of an electron at two points P and Q in a circular Bohr orbit

In the example shown in Figure 5.1, we define both the position and velocity in terms of vectors. The position of the electron at any given time is given by a position vector, referenced to an origin O. So, when the electron is at point P, its location is defined by the vector $\boldsymbol{a}$, whereas when it has

5.2 Addition and Subtraction of Vectors

5.2.1 Vector Addition

Consider the two vectors $\boldsymbol{a}$ and $\boldsymbol{b}$ shown in Figure 5.3:

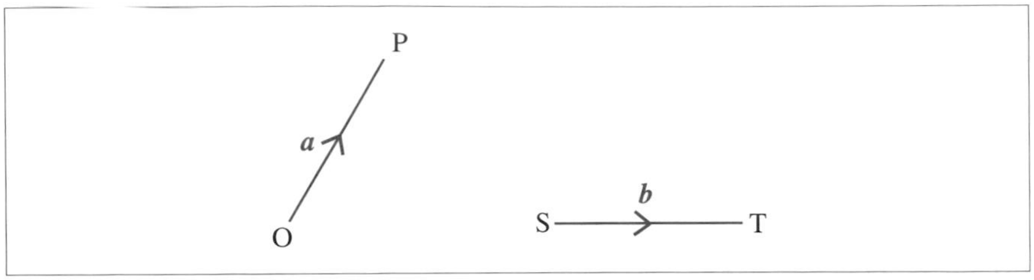

Figure 5.3 Vectors ***a*** and ***b*** with initial points O and S, respectively

The sum of $\boldsymbol{a}$ and $\boldsymbol{b}$ is given by the vector $\boldsymbol{c}$, which is found in the following series of steps:

(a) Translate the vector $\boldsymbol{b}$ until its initial point coincides with that of $\boldsymbol{a}$:

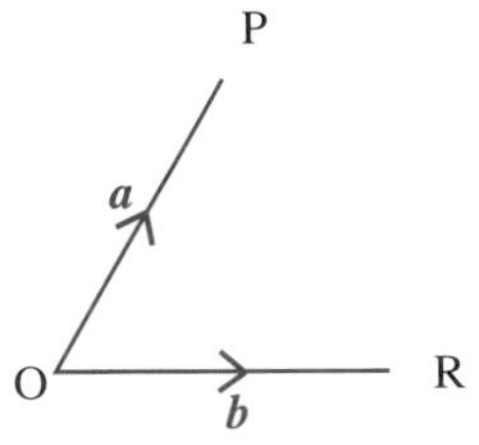

(b) Construct a parallelogram as indicated in Figure 5.4:

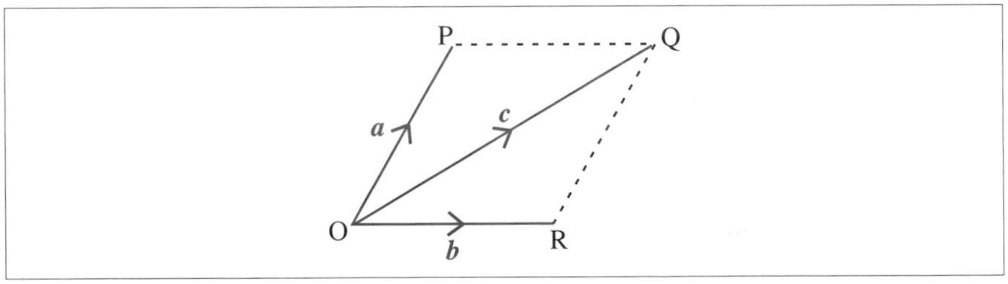

Figure 5.4 The parallelogram formed in the addition of the two vectors ***a*** and ***b***

The directed line segment $\overrightarrow{OQ}$ represents the vector $\boldsymbol{c}$, defined as the sum of $\boldsymbol{a}$ and $\boldsymbol{b}$. Furthermore, as $\overrightarrow{OQ} = \overrightarrow{OP} + \overrightarrow{PQ} = \overrightarrow{OR} + \overrightarrow{RQ}$, it follows that $\boldsymbol{c} = \boldsymbol{a} + \boldsymbol{b} = \boldsymbol{b} + \boldsymbol{a}$, from which we see that addition is **commutative**; in other words, a displacement $\overrightarrow{OR}$ followed by $\overrightarrow{RQ}$ clearly leads to the same final point as a displacement $\overrightarrow{OP}$ followed by $\overrightarrow{PQ}$.

Since $\overrightarrow{OR}$ and $\overrightarrow{PQ}$ are equivalent, and represent the same vector $\boldsymbol{b}$, we can use a triangle to summarize the relationship between $\boldsymbol{a}$, $\boldsymbol{b}$ and the resultant vector $\boldsymbol{c}$ (see Figure 5.5). For this reason, the equality $\boldsymbol{c} = \boldsymbol{a} + \boldsymbol{b}$ is often known as the **triangle rule**.

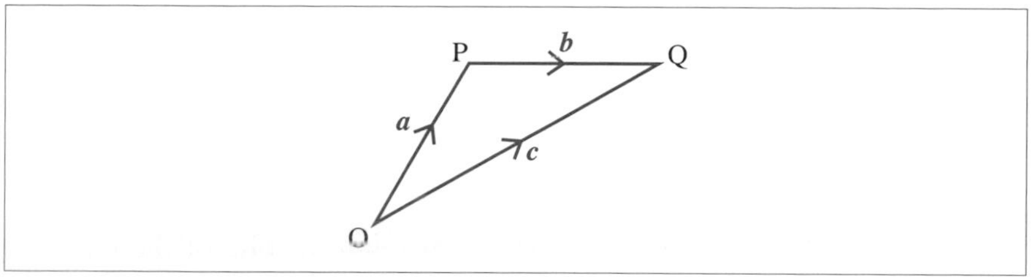

Figure 5.5 The triangle rule, in which $\boldsymbol{c} = \boldsymbol{a} + \boldsymbol{b}$

5.2.2 Vector Subtraction

The subtraction of two vectors can be thought of as the addition of two vectors that differ in their sign. If we think of this in terms of displacements in space, then the first vector corresponds to a displacement from point P to point Q, for example, whereas a second identical vector with opposite sign will direct us back to point P from point Q:

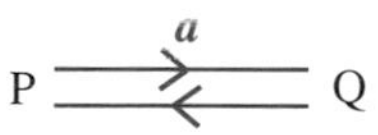

The net result is the null vector – we end up where we started:

$$\boldsymbol{a} + -\boldsymbol{a} = \boldsymbol{a} - \boldsymbol{a} = \boldsymbol{0} \tag{5.2}$$

It follows that subtraction of two vectors, $\boldsymbol{a}$ and $\boldsymbol{b}$, is equivalent to adding the vectors $\boldsymbol{a}$ and $-\boldsymbol{b}$, and so we can define vector subtraction in a general sense as:

$$\boldsymbol{d} = \boldsymbol{a} + (-\boldsymbol{b}) = \boldsymbol{a} - \boldsymbol{b} \tag{5.3}$$

which can be expressed in terms of a variant of the triangle rule, as seen in Figure 5.6.

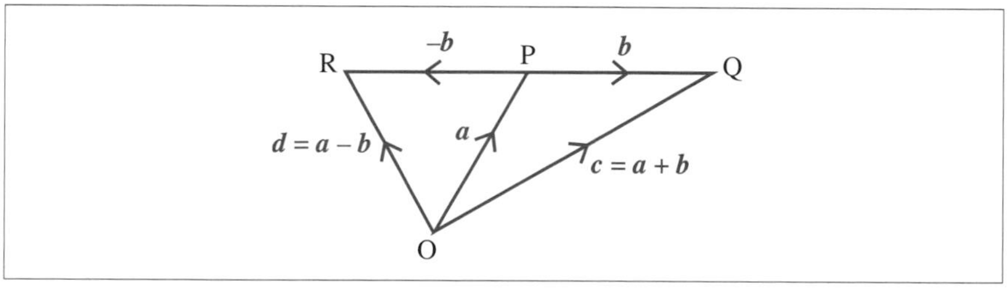

Figure 5.6 Vector subtraction represented in terms of the triangle rule

It also follows from Figure 5.6 that if $\boldsymbol{a} + \boldsymbol{b} = \boldsymbol{c}$, then $\boldsymbol{c} - \boldsymbol{a} = \boldsymbol{b}$, which we represent graphically in two ways in Figure 5.7:

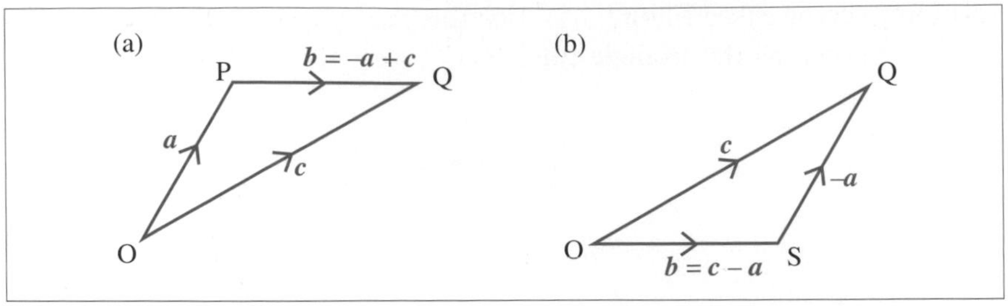

Figure 5.7 Two alternative representations of the subtraction of two vectors

Note that both representations are equivalent, in spite of the fact that the initial and final points of vector $\boldsymbol{b}$ are located at different points in space in the two representations. Thus, since the vector is fully defined simply by its direction and magnitude, the locations of the initial and final points are unimportant, unless we define them to act in specific locations.

Problem 5.1

Use the vectors $\boldsymbol{a}$ and $\boldsymbol{b}$ in Figure 5.5 to construct parallelograms, defined by the vectors: $\boldsymbol{c} = \boldsymbol{a} + 2\boldsymbol{b}$ and $\boldsymbol{d} = 2\boldsymbol{a} - \boldsymbol{b}$.

5.3 Base Vectors

Any kind of operation on a vector, including addition and subtraction, can be somewhat laborious when working with its graphical representation. However, by referring the vectors to a common set of unit vectors, termed **base vectors**, we can reduce the manipulations of vectors to algebraic operations.

In three-dimensional space, a convenient set of three unit vectors is provided by $\hat{\boldsymbol{i}}, \hat{\boldsymbol{j}}$ and $\hat{\boldsymbol{k}}$, which are directed along the x, y, and z Cartesian axes, respectively (Figure 5.8).

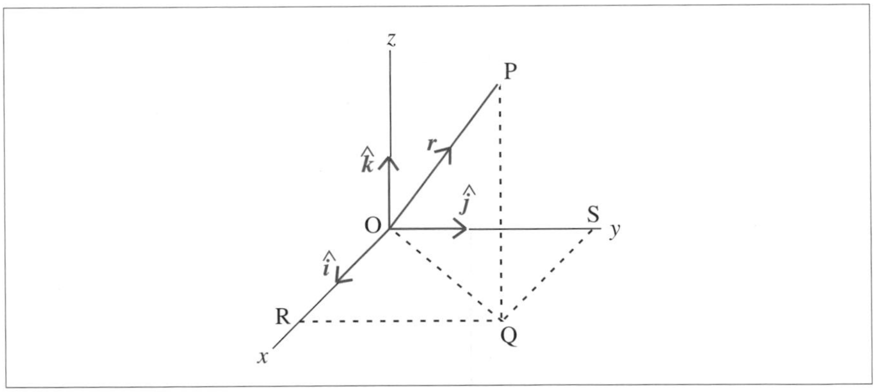

Figure 5.8 Base vectors in three dimensions for the Cartesian coordinate system

In this system of coordinates, if a point P has the coordinates (x,y,z), then the directed line segment $\overrightarrow{\mathrm{OP}}$, extending from the origin O to point P, corresponds to the vector $\boldsymbol{r}$. If we apply the triangle rule twice, we obtain:

$$\boldsymbol{r} = \overrightarrow{\mathrm{OQ}} + \overrightarrow{\mathrm{QP}} \tag{5.4}$$

$$= \overrightarrow{\mathrm{OR}} + \overrightarrow{\mathrm{RQ}} + \overrightarrow{\mathrm{QP}} \tag{5.5}$$

$$\Rightarrow \quad \boldsymbol{r} = x\hat{\boldsymbol{i}} + y\hat{\boldsymbol{j}} + z\hat{\boldsymbol{k}} \tag{5.6}$$

Equation (5.6) expresses $\boldsymbol{r}$ as a sum of the vectors $x\hat{\boldsymbol{i}}$, $y\hat{\boldsymbol{j}}$ and $z\hat{\boldsymbol{k}}$, which are called the **projections** of $\boldsymbol{r}$ in the direction of the x-, y- and z-axes. The magnitudes of each projection are given by the x-, y- and z-values, respectively, defining the location of P; however, in the context of vectors, these values (coordinates) are known as the **components** of $\boldsymbol{r}$; if the components are all zero, then this defines the null vector. Note that for problems in two dimensions, only two base vectors are required, such as, for example, $\hat{\boldsymbol{i}}$ and $\hat{\boldsymbol{j}}$.

The Magnitude of $\overrightarrow{\mathrm{OP}}$

If we apply the Pythagoras' theorem, first to triangle ORQ in Figure 5.8, and then to triangle OQP, we obtain an expression for the **magnitude** of $\boldsymbol{r}$ in terms of its components:

$$|\boldsymbol{r}| = (x^2 + y^2 + z^2)^{1/2} \tag{5.7}$$

5.3.1 Vector Addition, Subtraction and Scalar Multiplication using Algebra

The algebraic approach to vector addition and subtraction simply involves adding or subtracting the respective projections, $x\hat{\boldsymbol{i}}$, $y\hat{\boldsymbol{j}}$ and $z\hat{\boldsymbol{k}}$, of the two (or more) vectors. **Scalar multiplication** requires each projection to be multiplied by the scalar quantity.

Worked Problem 5.1

Q If $\boldsymbol{u} = \hat{\boldsymbol{i}} + \hat{\boldsymbol{j}} + 2\hat{\boldsymbol{k}}$ and $\boldsymbol{v} = -2\hat{\boldsymbol{i}} - \hat{\boldsymbol{j}} + \hat{\boldsymbol{k}}$, find: (a) $2\boldsymbol{v}$; (b) $\boldsymbol{u} - 2\boldsymbol{v}$; (c) $|\boldsymbol{u} - 2\boldsymbol{v}|$.

A (a) $2\boldsymbol{v} = -4\hat{\boldsymbol{i}} - 2\hat{\boldsymbol{j}} + 2\hat{\boldsymbol{k}}$.
(b) $\boldsymbol{u} - 2\boldsymbol{v} = (\hat{\boldsymbol{i}} + \hat{\boldsymbol{j}} + 2\hat{\boldsymbol{k}}) - (-4\hat{\boldsymbol{i}} - 2\hat{\boldsymbol{j}} + 2\hat{\boldsymbol{k}}) = (5\hat{\boldsymbol{i}} + 3\hat{\boldsymbol{j}})$.
(c) $|\boldsymbol{u} - 2\boldsymbol{v}| = \sqrt{5^2 + 3^2} = \sqrt{34}$.

Problem 5.2

If $\boldsymbol{a} = \hat{\imath} + \hat{\jmath} - 2\hat{k}$, $\boldsymbol{b} = \hat{\imath} + \hat{k}$, $\boldsymbol{c} = \hat{\imath} + \hat{\jmath} + \hat{k}$ and $\boldsymbol{d} = \hat{\imath} - 2\hat{k}$, find:
(a) $3\boldsymbol{a} - 2\boldsymbol{b}$; (b) $-2\boldsymbol{a} - \boldsymbol{b}$; (c) $\boldsymbol{a} + \boldsymbol{b} - \boldsymbol{c} - \boldsymbol{d}$; (d) $|\boldsymbol{a} - \boldsymbol{d}|$;
(e) $(\boldsymbol{a} + \boldsymbol{c})/|\boldsymbol{a} + \boldsymbol{c}|$; (f) the magnitude of the vector in (e);
(g) $|\boldsymbol{a}| - |\boldsymbol{c}|$.

Problem 5.3

Consider the planar complex ion $Co(CN)_4^{2-}$, shown schematically in Figure 5.9. The central Co lies at the origin, and the four CN^- ligands lie on either the x- or y- axis; R is the Co–C interatomic distance.

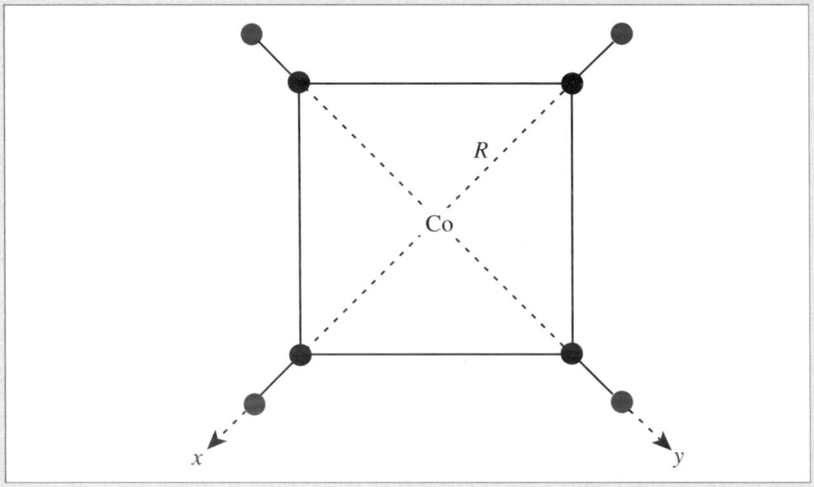

Figure 5.9 The planar complex ion $Co(CN)_4^{2-}$, where the carbon atoms are represented by *black spheres* and the nitrogen atoms by *coloured spheres*

(a) Identify the unit vectors directed toward each of the four CN^- ligands.

(b) Give the forms of the four vectors, directed from Co to each C atom.

(c) Find the vectors specifying one of the shortest and one of the longest C–C distances, and hence determine these distances in terms of R.

Hint: the representation of vector subtraction shown in Figure 5.7(a) may be helpful.

5.4 Multiplication of Vectors

In algebra, as we saw in Chapter 2 of Volume 1, the act of multiplication is an unambiguous and well-defined operation indicated by the sign $\times$. In the algebra of vectors, however, multiplication and division have no obvious conventional meaning. Despite this drawback, the two kinds of multiplication operation on pairs of vectors in widespread use are defined in the following subsections.

The $\times$ symbol used in the multiplication of numbers and symbols is commonly suppressed; thus, $6xy$ is shorthand for the product $6 \times x \times y$.

Objects formed by placing vectors in juxtaposition, such as ***ab***, are called dyadics and have a role in theoretical aspects of Raman spectroscopy, for example.

5.4.1 Scalar Product of Two Vectors

Consider the vectors $\boldsymbol{a}$ and $\boldsymbol{b}$ in Figure 5.10, in which the angle between the two vectors is θ:

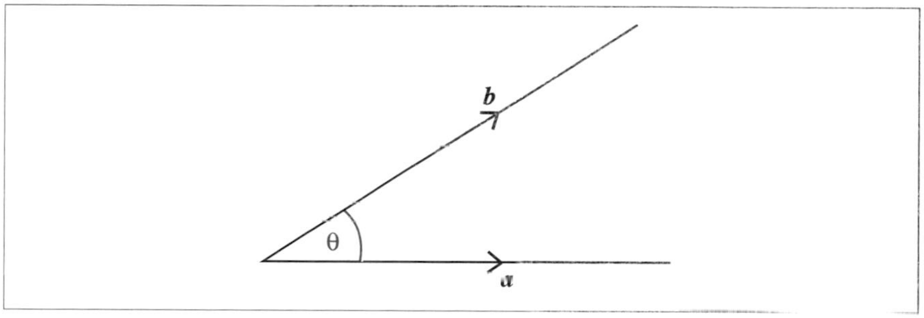

Figure 5.10 Two vectors ***a*** and ***b***, inclined with respect to one another at an angle θ

The scalar product is defined as:

$$\boldsymbol{a} \cdot \boldsymbol{b} = \boldsymbol{b} \cdot \boldsymbol{a} = |\boldsymbol{a}||\boldsymbol{b}|\cos\theta \tag{5.8}$$

The right side of equation (5.8) indicates that the result is a scalar (number), and *not* another vector, because it involves the product of the magnitudes of the two vectors, with the cosine of the angle between them (a positive or negative number, depending on the angle). Thus, since $|\boldsymbol{a}|$ and $|\boldsymbol{b}|$ are, by definition, positive numbers, the sign of the scalar product is determined by the value of the angle θ. In particular, the scalar product is:

- Positive for an acute angle ($\theta < 90°$).
- Zero for $\theta = 90°$.
- Negative for an obtuse angle ($90° < \theta \leqslant 180°$).

By convention, the angle θ is restricted to the range $0 \leqslant \theta \leqslant 180°$. If $\theta = 90°$, then $\boldsymbol{a}.\boldsymbol{b} = 0$, and $\boldsymbol{a}$ and $\boldsymbol{b}$ are said to be **orthogonal**. On the other hand, the scalar product of a vector with itself ($\theta = 0°$; $\cos\theta = 1$), yields the square of its magnitude:

$$\boldsymbol{a} \cdot \boldsymbol{a} = |\boldsymbol{a}|^2, \text{ implying that } |\boldsymbol{a}| = \sqrt{\boldsymbol{a} \cdot \boldsymbol{a}} \tag{5.9}$$

Furthermore, if $\boldsymbol{a}$ is of unit magnitude, then $\boldsymbol{a} \cdot \boldsymbol{a} = 1$, and $\boldsymbol{a}$ is said to be **normalized**.

Specifying the Angle θ

In some situations, it is important to be aware of how the sense of direction of the two vectors $\boldsymbol{a}$ and $\boldsymbol{b}$ affects the choice for the value of the angle between them. For example, the angle θ between the vectors in Figure 5.10 constitutes the correct choice, because the two vectors are directed away from the common point of origin. However, if vector $\boldsymbol{a}$ (or vector $\boldsymbol{b}$) is directed in the opposite sense (the dashed directed line segment in Figure 5.11), then we determine the angle between $\boldsymbol{a}$ and $\boldsymbol{b}$ by realigning the two vectors to ensure once again that they are directed away from the common origin point. The angle is then defined as $180° - \theta$ (Figure 5.11).

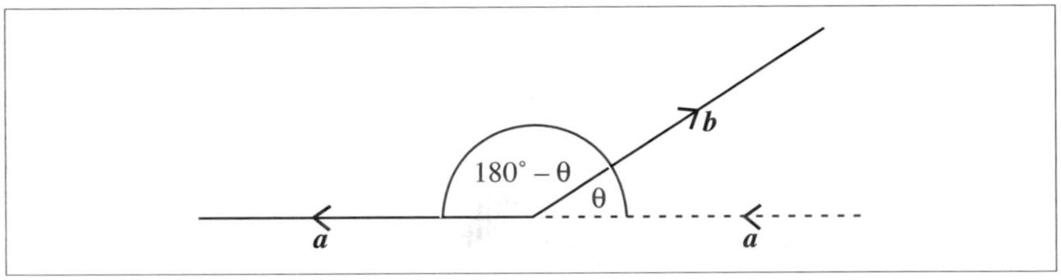

Figure 5.11 The relationship between vector direction and angle

The Scalar Product in the Chemical Context

Scalar products arise in a number of important areas in chemistry. For example, they are involved in:

- Determining the energy, W, of a molecular electric or magnetic dipole interacting with an electric or magnetic field, $W = -\boldsymbol{\mu}_e \cdot \mathbf{E}$ or $W = -\boldsymbol{\mu}_m \cdot \mathbf{H}$, respectively.
- Evaluating the consequences of the intermolecular dipole–dipole interactions in molecular crystalline solids.
- Crystallography, where the **scalar triple product** (see Section 5.5.3) is used to evaluate the volume of a crystallographic unit cell.

Scalar Products of Vectors Expressed in Terms of Base Vectors

The scalar product of two vectors $\boldsymbol{a}$ and $\boldsymbol{b}$, expressed in terms of base vectors, is obtained by taking the sum of the scalar products of each base vector pair, together with the appropriate product of components.

Worked Problem 5.2

Q Find the scalar product of the vectors $\boldsymbol{a} = \hat{\imath} + \hat{\jmath} - 2\hat{k}$ and $\boldsymbol{b} = \hat{\imath} + \hat{k}$.

A We find the scalar product of $\boldsymbol{a}$ and $\boldsymbol{b}$ using the respective components (1,1,−2) and (1,0,1). Thus, expanding the brackets yields:

$$\begin{aligned} \boldsymbol{a} \cdot \boldsymbol{b} &= (\hat{\imath} + \hat{\jmath} - 2\hat{k}) \cdot (\hat{\imath} + \hat{k}) \\ &= \hat{\imath} \cdot \hat{\imath} + \hat{\imath} \cdot \hat{k} + \hat{\jmath} \cdot \hat{\imath} + \hat{\jmath} \cdot \hat{k} - 2\hat{k} \cdot \hat{\imath} - 2\hat{k} \cdot \hat{k} \end{aligned} \quad (5.10)$$

We now use equation (5.8) to evaluate each scalar product of basc vectors, to obtain:

$$\hat{\imath} \cdot \hat{\imath} = \hat{\jmath} \cdot \hat{\jmath} = \hat{k} \cdot \hat{k} = 1 \quad (\theta = 0) \quad (5.11)$$

$$\hat{\imath} \cdot \hat{\jmath} = \hat{\imath} \cdot \hat{k} = \hat{k} \cdot \hat{\jmath} = 0 \quad (\theta = 90^\circ) \quad (5.12)$$

which, on substitution into equation (5.10), gives:

$$\begin{aligned} \boldsymbol{a} \cdot \boldsymbol{b} &= 1 + 0 + 0 + 0 - 0 - 2 \\ &= -1 \end{aligned}$$

Problem 5.4

If $\boldsymbol{a} = 2\hat{\imath} + 3\hat{k}$, $\boldsymbol{b} = \hat{\imath} + \hat{\jmath} + \hat{k}$ and $\boldsymbol{c} = \hat{\imath} - 2\hat{\jmath} + \hat{k}$, find: (a) $\boldsymbol{a} \cdot \boldsymbol{c}$; (b) $\boldsymbol{a} \cdot (\boldsymbol{b} - 2\boldsymbol{c})$; (c) $\boldsymbol{a} \cdot (\boldsymbol{b} + \boldsymbol{a})$; (d) $\boldsymbol{b} \cdot \boldsymbol{c}$.

Finding the Angle Between Two Vectors

In the previous section we saw that, in spite of appearances, we do not need to know the angle between two vectors in order to evaluate the scalar product according to equation (5.8): we simply exploit the properties of the **orthonormal** base vectors to evaluate the result algebraically. However, we can approach from a different perspective, and use the right side of equation (5.8) to find the angle between two vectors, having evaluated the scalar product using the approach detailed above. The next Worked Problem details how this is accomplished.

Vectors which are orthogonal to one another, as well as being normalized, are said to be orthonormal.

Worked Problem 5.3

Q Use equation (5.8) to find the angle between the vectors $\boldsymbol{a} = \hat{\imath} + \hat{\jmath} - 2\hat{k}$ and $\boldsymbol{b} = \hat{\imath} + \hat{k}$.

A Since the scalar product of these two vectors is negative, and has the value −1 (Worked Problem 5.2), we know that the angle θ is obtuse. The next step involves substitution of the vector magnitudes $|\boldsymbol{a}| = \sqrt{6}$ and $|\boldsymbol{b}|\sqrt{2}$ into equation (5.8), in order to determine the value of cos θ:

$$-1 = \sqrt{6} \times \sqrt{2} \times \cos\theta \Rightarrow \cos\theta = -\frac{1}{\sqrt{2}\sqrt{6}} = -\frac{1}{2\sqrt{3}}.$$

Since cos θ is negative (an obtuse angle), and θ is restricted to $0 \leqslant \theta \leqslant 180°$, we obtain the result $\theta = 106.7°$.

Problem 5.5

(a) Using the definition of the vectors $\boldsymbol{a}$, $\boldsymbol{b}$, and $\boldsymbol{c}$ in Problem 5.4, find the angle between (i) $\boldsymbol{a}$ and $(\boldsymbol{b}-2\boldsymbol{c})$ and (ii) $\boldsymbol{b}$ and $\boldsymbol{c}$.
(b) Find a value of λ for which the two vectors $\boldsymbol{d} = 3\hat{\boldsymbol{i}} - 2\hat{\boldsymbol{j}} - \hat{\boldsymbol{k}}$ and $\boldsymbol{e} = \hat{\boldsymbol{i}} + \lambda\hat{\boldsymbol{j}} + 2\hat{\boldsymbol{k}}$ are orthogonal.

Simple Application of the Scalar Product: the Cosine Law for a Triangle

If the sides of the triangle OPQ in Figure 5.5, formed from the vectors $\boldsymbol{a}$, $\boldsymbol{b}$ and $\boldsymbol{c}$ have magnitudes a, b and c, respectively, and B is the angle opposite $\boldsymbol{b}$, then we can use equations (5.8) and (5.9) to find a useful relationship between a, b, c and B.

The triangle rule $\boldsymbol{c} = \boldsymbol{a} + \boldsymbol{b}$ may be rewritten as $\boldsymbol{b} = \boldsymbol{c} - \boldsymbol{a}$, from which we form the scalar product $\boldsymbol{b} \cdot \boldsymbol{b}$:

$$\boldsymbol{b} \cdot \boldsymbol{b} = (\boldsymbol{c} - \boldsymbol{a}) \cdot (\boldsymbol{c} - \boldsymbol{a}) = \boldsymbol{a} \cdot \boldsymbol{a} + \boldsymbol{c} \cdot \boldsymbol{c} - 2\boldsymbol{a} \cdot \boldsymbol{c} \qquad (5.13)$$

However, since we know that the scalar product of a vector with itself yields the square of its magnitude (equation 5.9), and that the angle between $\boldsymbol{a}$ and $\boldsymbol{c}$ is B (and not $180° - B$), it follows that:

$$\boldsymbol{b} \cdot \boldsymbol{b} = b^2 = a^2 + c^2 - 2ac\cos B \qquad (5.14)$$

This can be extended to construct analogous expressions involving the angles A and C, opposite vectors $\boldsymbol{a}$ and $\boldsymbol{c}$, respectively.

Problem 5.6

Use the triangle rule in the form $\boldsymbol{c} = \boldsymbol{a} + \boldsymbol{b}$ to derive the form of the cosine formula involving the angle C.

Hint: find an expression for $\boldsymbol{c}\cdot\boldsymbol{c}$ and decide, using Figure 5.11, whether the angle θ (in degrees) between the vectors $\boldsymbol{a}$ and $\boldsymbol{b}$ is the same as angle C or the angle $180° - C$. *Remember*: the cosine formula given above always has the same form, regardless of the choice of a, b and c or angles A, B and C.

Problem 5.7

The complex ion $CoCl_4^{2-}$ adopts a tetrahedral shape, in which the Co lies at the centre of a cube of side $2a$, and the Cl^- species are located on alternate cube vertices; the Co–Cl interatomic distance is taken as R. The coordinate axes are chosen to pass through the centres of opposite pairs of cube faces, with the Co lying at the origin, as shown in Figure 5.12.

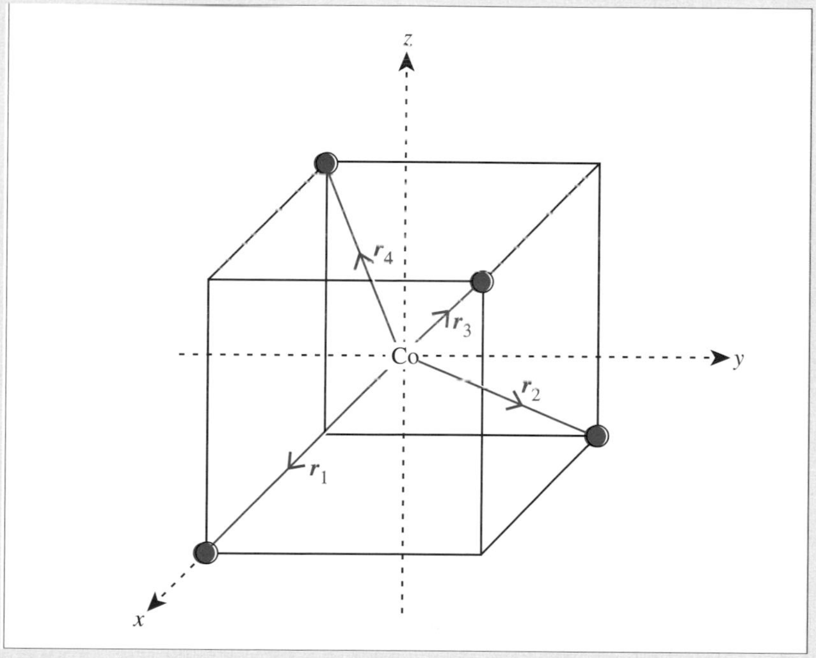

Figure 5.12 The complex ion $CoCl_4^{2-}$, where the *coloured spheres* represent a Cl^- species

(a) Given that the coordinates of the four Cl^- ligands are $(a,-a,-a)$, $(-a,a,-a)$, (a,a,a) and $(-a,-a,a)$, write down the algebraic form of the four vectors, $\boldsymbol{r}_1$, $\boldsymbol{r}_2$, $\boldsymbol{r}_3$ and $\boldsymbol{r}_4$, directed from the central Co (0,0,0) to the four ligands.

(b) Find the magnitude of any one of the Co–Cl vectors, and hence express a in terms of R.
(c) Use the triangle rule shown in Figure 5.7(a) to find a vector associated with the interligand distance, and hence find its magnitude in terms of R.

5.4.2 Vector Product of Two Vectors

In the previous section we defined the scalar product as a vector operation resembling the act of multiplication, which results in a scalar (or number, with or without units). We can now define a second type of vector multiplication known as the **vector product**, which results in another vector rather than a number. The vector product is defined as:

$$\boldsymbol{a} \times \boldsymbol{b} = -\boldsymbol{b} \times \boldsymbol{a} = |\boldsymbol{a}||\boldsymbol{b}| \sin \theta \cdot \hat{\boldsymbol{n}} \tag{5.15}$$

where $\hat{\boldsymbol{n}}$ is the unit vector orthogonal (perpendicular in 2-D or 3-D space) to the plane containing $\boldsymbol{a}$ and $\boldsymbol{b}$. Since there are two possible choices for $\hat{\boldsymbol{n}}$ (up or down), the convention for selecting the appropriate direction for $\hat{\boldsymbol{n}}$ requires the vectors $\boldsymbol{a}$, $\boldsymbol{b}$ and $\hat{\boldsymbol{n}}$ to form a right-handed system of axes, as shown in Figure 5.13.

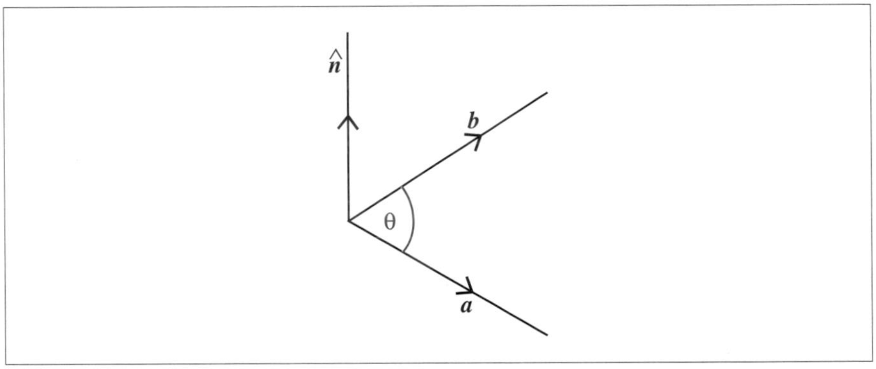

Figure 5.13 The axis convention for determining the sign of the unit vector $\hat{\boldsymbol{n}}$ directed perpendicular to the plane containing the vectors $\boldsymbol{a}$ and $\boldsymbol{b}$

If we imagine the action of a right-hand corkscrew, in which $\boldsymbol{a}$ is rotated towards $\boldsymbol{b}$, in an anticlockwise sense when viewed from above, the corkscrew moves in the direction of $\hat{\boldsymbol{n}}$; it follows that the analogous corkscrew motion taking $\boldsymbol{b}$ to $\boldsymbol{a}$ (clockwise) yields a movement in the direction of $-\hat{\boldsymbol{n}}$. Consequently, the vector products involving the base vectors $\hat{\boldsymbol{j}}$ and $\hat{\boldsymbol{k}}$, or, by suitable changes, any other pair of base vectors, are determined by forming a right-hand clockwise system, as seen in Figure 5.14.

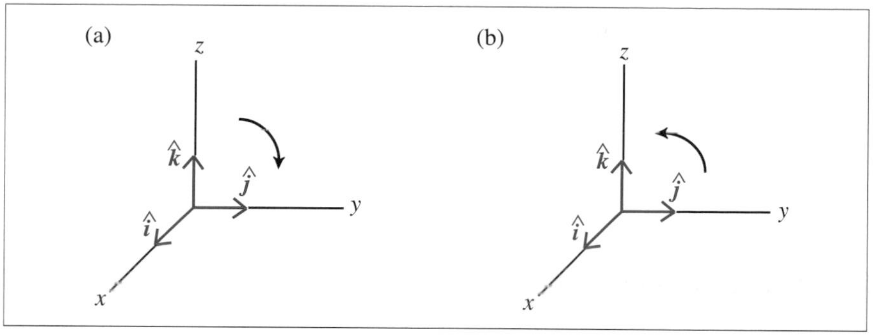

Figure 5.14 Formation of the vector product (a) $\hat{k} \times \hat{j} = -\hat{i}$ and (b) $\hat{j} \times \hat{k} = \hat{i}$ determined by looking down the *x*-axis and imagining the action of a right-handed corkscrew motion (see text for details)

Although analogous results can be derived for other pairs of base vectors, the simplest aid for obtaining the appropriate result is to use the diagram shown in Figure 5.15. The vector product $\hat{i} \times \hat{j}$, for example, is verified by moving in a clockwise manner from $\hat{i}$ to $\hat{j}$ to the next base vector $\hat{k}$. However, $\hat{j} \times \hat{i}$ yields $-\hat{k}$ because anticlockwise circulation introduces a negative sign.

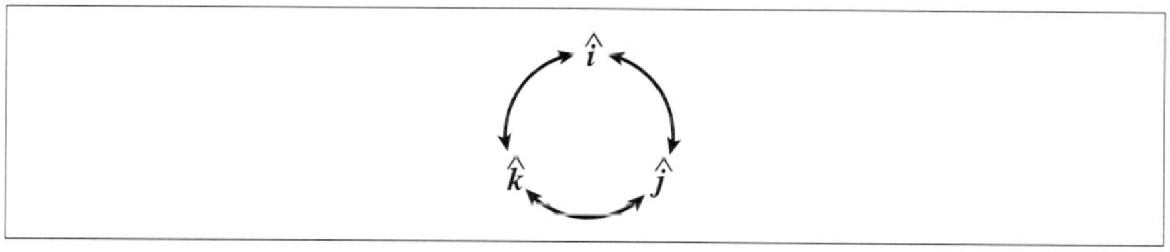

Figure 5.15 The vector product of any two base vectors, moving in a clockwise or anticlockwise direction, yields resultant vectors of positive or negative signs, respectively

In forming the vector product of two vectors $\boldsymbol{a}$ and $\boldsymbol{b}$, we should remember that:

- The order of operation is very important: the operation is not commutative (equation 5.15).
- The resulting vector is orthogonal to both $\boldsymbol{a}$ and $\boldsymbol{b}$, implying that:

$$\boldsymbol{a} \cdot (\boldsymbol{a} \times \boldsymbol{b}) = 0 \quad \text{and} \quad \boldsymbol{b} \cdot (\boldsymbol{a} \times \boldsymbol{b}) = 0$$

- The operation is not generally associative:

$$(\boldsymbol{a} \times \boldsymbol{b}) \times \boldsymbol{c} \neq \boldsymbol{a} \times (\boldsymbol{b} \times \boldsymbol{c})$$

Although the vector product is not generally associative, there are examples where this rule is violated – one being the three vectors given in Problem 5.11.

Vector Products in a Chemical Context

Vector products arise when:

- Working with the angular momentum, l (a vector property), associated with the circular motion of a particle of mass, m, moving under a constant potential about a fixed point with velocity and position described by the vectors:

$$\boldsymbol{v} = \begin{pmatrix} v_x \\ v_y \\ v_z \end{pmatrix} \text{ and } \boldsymbol{x} = \begin{pmatrix} x \\ y \\ z \end{pmatrix}$$

In this instance the angular momentum $\boldsymbol{l} = \boldsymbol{r} \times m\boldsymbol{v} = \boldsymbol{r} \times \boldsymbol{p}$, where

$$\boldsymbol{p} = \begin{pmatrix} mv_x \\ mv_y \\ mv_z \end{pmatrix}$$

is the linear momentum. Such model systems have particular relevance when considering the orbital motion of an electron about a nucleus in an atom, or about the internuclear axis in a linear molecule.

- Evaluating the volume of a crystallographic unit cell through the scalar triple product (see Section 5.5.3).

Worked Problem 5.4

Q (a) Use equation (5.15) to express the vector products $\hat{\boldsymbol{i}}\times\hat{\boldsymbol{i}}$, $\hat{\boldsymbol{j}}\times\hat{\boldsymbol{j}}$ and $\hat{\boldsymbol{k}}\times\hat{\boldsymbol{k}}$ in terms of the base vectors $\hat{\boldsymbol{i}}$, $\hat{\boldsymbol{j}}$ and $\hat{\boldsymbol{k}}$.

(b) With the aid of Figure 5.15, find the vector products $\hat{\boldsymbol{i}}\times\hat{\boldsymbol{j}}$, $\hat{\boldsymbol{i}}\times\hat{\boldsymbol{k}}$ and $\hat{\boldsymbol{j}}\times\hat{\boldsymbol{k}}$.

A (a) $\hat{\boldsymbol{i}}\times\hat{\boldsymbol{i}} = |\hat{\boldsymbol{i}}||\hat{\boldsymbol{i}}|\sin\theta\cdot\hat{\boldsymbol{n}} = 1\times 1\cdot\sin 0\cdot\hat{\boldsymbol{n}} = \boldsymbol{0}$, since $\sin 0 = 0$. The outcome is the same for $\hat{\boldsymbol{j}}\times\hat{\boldsymbol{j}}$ and $\hat{\boldsymbol{k}}\times\hat{\boldsymbol{k}}$, and so:

$$\hat{\boldsymbol{i}}\times\hat{\boldsymbol{i}} = \hat{\boldsymbol{j}}\times\hat{\boldsymbol{j}} = \hat{\boldsymbol{k}}\times\hat{\boldsymbol{k}} = \boldsymbol{0}$$

It is important not to confuse this result with the analogous scalar products.

(b) $\hat{\boldsymbol{i}}\times\hat{\boldsymbol{j}} = \hat{\boldsymbol{k}}$, $\hat{\boldsymbol{i}}\times\hat{\boldsymbol{k}} = -\hat{\boldsymbol{j}}$, $\hat{\boldsymbol{j}}\times\hat{\boldsymbol{k}} = \hat{\boldsymbol{i}}$.

Problem 5.8

Use the definitions of $\boldsymbol{a}$ and $\boldsymbol{c}$ in Problem 5.4, and the results of Worked Problem 5.4, to find: (a) $\boldsymbol{a}\times\boldsymbol{c}$; (b) $\boldsymbol{c}\times\boldsymbol{a}$; (c) $|\boldsymbol{c}\times\boldsymbol{a}|$; (d) $(\hat{\boldsymbol{i}}\times\hat{\boldsymbol{j}})\times\hat{\boldsymbol{j}}$; (e) $\hat{\boldsymbol{i}}\times(\hat{\boldsymbol{j}}\times\hat{\boldsymbol{j}})$.

5.4.3 Area of a Parallelogram

The vector product of $\boldsymbol{a}$ and $\boldsymbol{b}$ provides a route for calculating the area of a parallelogram. We explore this method in Worked Problem 5.5.

Worked Problem 5.5

Consider the parallelogram OPQR in shown in Figure 5.16. If we extend OR to point T and drop perpendicular lines from P to S and from Q to T, we construct a rectangle with the same area as the original parallelogram: a result achieved by chopping off the triangle OPS from the left side of the parallelogram and reattaching it at the right side.

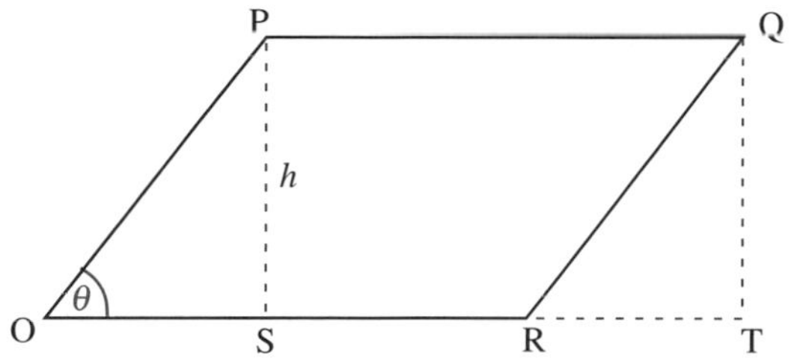

Figure 5.16 The area of the parallelogram OPQR is given by $|\boldsymbol{a} \times \boldsymbol{b}|$, where the vectors $\boldsymbol{a}$ and $\boldsymbol{b}$ represent the directed line segments $\overrightarrow{OR}$ and $\overrightarrow{OP}$, respectively

Q (a) Explain why the areas of the triangles OPS and RQT are equal.
(b) If the directed line segments $\overrightarrow{OR}$ and $\overrightarrow{OP}$, are represented by the vectors $\boldsymbol{a}$ and $\boldsymbol{b}$, show that the area of the parallelogram is given by $|\boldsymbol{a}|h$.
(c) Deduce that the area, A, of the parallelogram is given by $|\boldsymbol{a} \times \boldsymbol{b}|$.

A (a) The areas of the triangles OPS and RQT are equal because the lengths of the sides OP and RQ are the same, as are the angles ∠POS and ∠QRT.
(b) Given the equivalency of the two triangles OPS and RQT, the area of the rectangle SPQT must be the same as that of the parallelogram OPQR. Consequently the area of the parallelogram must be equal to the magnitude of $\overrightarrow{ST}$ multiplied by the height of the rectangle, h. Since $\overrightarrow{ST}$ has the same magnitude as $\overrightarrow{OR}$, and $\overrightarrow{OR}$ is represented by the vector $\boldsymbol{a}$, it follows that the magnitude of $\overrightarrow{ST}$ is equal to the magnitude of $\boldsymbol{a}$. Thus, the area $A = |\boldsymbol{a}|h$.
(c) As triangle OPS is right-angled, it follows that $h = |\boldsymbol{b}|\sin\theta$ and $A = |\boldsymbol{a}||\boldsymbol{b}|\sin\theta = |\boldsymbol{a} \times \boldsymbol{b}|$, where $\boldsymbol{a} \times \boldsymbol{b} = |\boldsymbol{a}||\boldsymbol{b}|\sin\theta \cdot \hat{n}$.

Problem 5.9

For $\boldsymbol{a} = a_1\hat{\boldsymbol{i}} + a_2\hat{\boldsymbol{j}} + a_3\hat{\boldsymbol{k}}$, $\boldsymbol{b} = b_1\hat{\boldsymbol{i}} + b_2\hat{\boldsymbol{j}} + b_3\hat{\boldsymbol{k}}$ and $\boldsymbol{c} = c_1\hat{\boldsymbol{i}} + c_2\hat{\boldsymbol{j}} + c_3\hat{\boldsymbol{k}}$, show that:

(a) $\boldsymbol{a} \cdot \boldsymbol{b} = a_1b_1 + a_2b_2 + a_3b_3$; (b) $\boldsymbol{b} \times \boldsymbol{c} = (b_2c_3 - b_3c_2)\,\hat{\boldsymbol{i}} - (b_1c_3 - b_3c_1)\,\hat{\boldsymbol{j}} + (b_1c_2 - b_2c_1)\hat{\boldsymbol{k}}$.

5.5 Matrices and Determinants Revisited: Alternative Routes to Determining Scalar and Vector Products

5.5.1 The Scalar Product

If the components of the vectors $\boldsymbol{a}$ and $\boldsymbol{b}$ in Problem 5.9 form the elements of the column matrices $\boldsymbol{v}_1 = \begin{pmatrix} a_1 \\ a_2 \\ a_3 \end{pmatrix}$ and $\boldsymbol{v}_2 = \begin{pmatrix} b_1 \\ b_2 \\ b_3 \end{pmatrix}$, then the scalar product $\boldsymbol{a} \cdot \boldsymbol{b}$ takes the form:

$$\boldsymbol{v}_1^{\mathrm{T}}\boldsymbol{v}_2 = \begin{pmatrix} a_1 & a_2 & a_3 \end{pmatrix}\begin{pmatrix} b_1 \\ b_2 \\ b_3 \end{pmatrix} = a_1b_1 + a_2b_2 + a_3b_3,$$

giving the same result as in Problem 5.9(a).

5.5.2 The Vector Product

If we compare the form of the vector product given in the answer to Problem 5.9(b) with the expansion of a determinant of order three, given in equation (3.20), we see that, if the correspondences:

$$\begin{array}{lll} a_{11} = \hat{\boldsymbol{i}}, & a_{12} = \hat{\boldsymbol{j}}, & a_{13} = \hat{\boldsymbol{k}} \\ a_{21} = b_1, & a_{22} = b_2, & a_{23} = b_3 \\ a_{31} = c_1, & a_{32} = c_2, & a_{33} = c_3 \end{array}$$

are made, then:

$$\boldsymbol{b} \times \boldsymbol{c} = \begin{vmatrix} \hat{\boldsymbol{i}} & \hat{\boldsymbol{j}} & \hat{\boldsymbol{k}} \\ b_1 & b_2 & b_3 \\ c_1 & c_2 & c_3 \end{vmatrix} \tag{5.16}$$

Using the properties of determinants, we see that exchanging rows two and three results in a change of sign of the determinant: such a change corresponds to the vector product $\boldsymbol{c} \times \boldsymbol{b}$ and is consistent with equation (5.15) and $\boldsymbol{b} \times \boldsymbol{c} = -\boldsymbol{c} \times \boldsymbol{b}$.

Problem 5.10

If $\boldsymbol{a}=\hat{\boldsymbol{i}}+\hat{\boldsymbol{j}}+\hat{\boldsymbol{k}}$ and $\boldsymbol{b}=\hat{\boldsymbol{i}}-\hat{\boldsymbol{j}}+\hat{\boldsymbol{k}}$: (a) use equation (5.16) to find the vector $\boldsymbol{a}\times\boldsymbol{b}$; (b) find $|\boldsymbol{a}\times\boldsymbol{b}|$ and specify a unit vector in the direction of $\boldsymbol{a}\times\boldsymbol{b}$.

5.5.3 The Scalar Triple Product

If we define three vectors $\boldsymbol{a}$, $\boldsymbol{b}$ and $\boldsymbol{c}$, as in Problem 5.9, the expression $\boldsymbol{a}\cdot(\boldsymbol{b}\times\boldsymbol{c})$, known as the **scalar triple product**, yields a scalar quantity, the magnitude of which provides the formula for the volume, V, of a parallelepiped with adjacent edges defined by vectors $\boldsymbol{a}$, $\boldsymbol{b}$ and $\boldsymbol{c}$ (an example in chemistry being a crystalline unit cell). If the determinantal representation of $\boldsymbol{b}\times\boldsymbol{c}$ is used, then, on expanding the determinant from the first row, and evaluating the three scalar products, we obtain:

$$\boldsymbol{a}\cdot(\boldsymbol{b}\times\boldsymbol{c})=\boldsymbol{a}\cdot\begin{vmatrix}\hat{\boldsymbol{i}} & \hat{\boldsymbol{j}} & \hat{\boldsymbol{k}}\\ b_1 & b_2 & b_3\\ c_1 & c_2 & c_3\end{vmatrix}$$

$$=\boldsymbol{a}\cdot\left\{\hat{\boldsymbol{i}}(b_2c_3-b_3c_2)-\hat{\boldsymbol{j}}\,(b_1c_3-b_3c_1)+\hat{\boldsymbol{k}}\,(b_1c_2-b_2c_1)\right\} \quad (5.17)$$

$$=a_1(b_2c_3-b_3c_2)-a_2(b_1c_3-b_3c_1)+a_3(b_1c_2-b_2c_1) \quad (5.18)$$

which, in turn, may be converted back into determinantal form:

$$\boldsymbol{a}\cdot(\boldsymbol{b}\times\boldsymbol{c})=\begin{vmatrix}a_1 & a_2 & a_3\\ b_1 & b_2 & b_3\\ c_1 & c_2 & c_3\end{vmatrix} \quad (5.19)$$

Thus, the volume (a positive quantity), V, of the parallelepiped formed from $\boldsymbol{a}$, $\boldsymbol{b}$ and $\boldsymbol{c}$ has the formula:

$$V=|\boldsymbol{a}\cdot(\boldsymbol{b}\times\boldsymbol{c})| \quad (5.20)$$

Since $\boldsymbol{a}\cdot(\boldsymbol{b}\times\boldsymbol{c})$ may be negative, we take the modulus to ensure a positive result.

We explore the application of the scalar triple product for evaluating the volume of a crystallographic unit cell in the final two problems of this chapter.

Problem 5.11

If $\boldsymbol{a}=a_2\,\hat{\boldsymbol{j}}+a_3\hat{\boldsymbol{k}}$, $\boldsymbol{b}=b_1\hat{\boldsymbol{i}}$ and $\boldsymbol{c}=c_2\hat{\boldsymbol{j}}$, use equation (5.19) to find an expression for $\boldsymbol{a}\cdot(\boldsymbol{b}\times\boldsymbol{c})$.

Problem 5.12

Crystalline naphthalene has a monoclinic unit cell, defined by the vectors in Problem 5.11, where $|\boldsymbol{a}| = 0.824$ nm, $|\boldsymbol{b}| = 0.600$ nm, $|\boldsymbol{c}| = 0.866$ nm and the angles between $\boldsymbol{a}$ and $\boldsymbol{b}$, $\boldsymbol{a}$ and $\boldsymbol{c}$, and $\boldsymbol{b}$ and $\boldsymbol{c}$ are $\alpha = 90°$, $\beta = 122.9°$ and $\gamma = 90°$, respectively:

(a) Give the values of b_1 and c_2.

(b) Use equation (5.8) for the scalar product $\boldsymbol{a} \cdot \boldsymbol{c}$ and your answer to Problem 5.11 to show that $a_2 \times 0.866\ \text{nm} = 0.824 \times 0.866 \times \cos\beta\ \text{nm}^2$, and hence find the value of a_2.

(c) Use equation (5.9) to show that $|\boldsymbol{a}| = \sqrt{a_2^2 + a_3^2}$, and hence find the two possible values for a_3.

(d) Use equation (5.19) to calculate the volume of the unit cell for naphthalene, using the positive value for a_3 obtained in (c).

Note: repeating the calculation of the volume of the unit cell using the negative value for a_3 yields an identical result for the volume of the unit cell. The negative value for a_3 arises as a legitimate mathematical solution, but has little physical relevance other than to reflect the unit cell in the xy-plane.

Summary of Key Points

This chapter provides a description of some of the mathematical tools required to understand the properties of chemical and physical quantities that are defined not only by magnitude but also by direction. The key points discussed include:

1. The graphical definition of a vector.
2. A geometrical method for the addition and subtraction of vectors.
3. The properties of Cartesian base vectors.
4. An algebraic method for the addition and subtraction of vectors.
5. The scalar and vector products of two vectors.
6. The scalar triple products, involving three vectors.
7. Working with vectors using matrix and determinantal notation.
8. A selection of mathematical and chemically based examples, to illustrate practical applications of vectors.

Answers to Problems

Chapter 1

1.1. (a) $\lim\limits_{r\to\infty} \frac{1}{2^r} = 0$; converges.

(b) $\lim\limits_{n\to\infty} \dfrac{n-1}{2n} = \dfrac{n}{2n} = \dfrac{1}{2}$; converges.

(c) $\lim\limits_{r\to\infty} \cos r\pi$; oscillates between ± 1.

1.2. $n = 6$; $r = 0, 1, 2, 3, 4, 5, 6$; ${}^6C_r = \frac{6!}{(6-r)!r!}$

$$\Rightarrow \frac{720}{720\times 1} = 1;\ \frac{720}{120\times 1} = 6;\ \frac{720}{24\times 2} = 15;\ \frac{720}{6\times 6} = 20;$$

$$\frac{720}{2\times 24} = 15;\ \frac{720}{1\times 120} = 6;\ \frac{720}{1\times 720} = 1.$$

1.3. Geometric series $1, 2, 4, 8, \ldots 2^r$, $\Rightarrow a = 1$, $ax = 2$, $ax^2 = 4$, $\Rightarrow$ $a = 1$, $x = 2$.
Using equation (1.20): $S_n = a\left(\dfrac{1-x^n}{1-x}\right) = \dfrac{1-2^n}{1-2} = \dfrac{1-2^n}{-1} = 2^n - 1.$

1.4. (a) $S = 1 + 2x + 3x^2 + 4x^3 + \cdots$:

$$u_r = rx^{r-1};\ \frac{u_{r+1}}{u_r} = \frac{(r+1)x^r}{rx^{r-1}} = \left(\frac{r+1}{r}\right)x$$

$$\therefore \lim_{r\to\infty}\left|\frac{u_{r+1}}{u_r}\right| = \lim_{r\to\infty}\left|\left(\frac{r+1}{r}\right)x\right| = |x|;\ \text{converges if}$$

$$|x| < 1,\ \textit{i.e.}\ \text{for}\ -1 < x < 1.$$

(b) $S = 1 - x + \frac{x^2}{2!} - \frac{x^3}{3!} + \frac{x^4}{4!} - \ldots + (-1)^{r-1}\frac{x^{r-1}}{(r-1)!} + \cdots$:

$$\frac{u_{r+1}}{u_r} = \frac{(-1)^r \frac{x^r}{r!}}{(-1)^{r-1}\frac{x^{r-1}}{(r-1)!}} = \frac{(-1)(r-1)!x^r}{r!x^{r-1}} = \frac{-x}{r}$$

$$\lim_{r\to\infty}\left|\frac{u_{r+1}}{u_r}\right| = \left|\frac{-x}{r}\right| = 0, \therefore \text{ converges for all } x.$$

(c) $S = 1 + \frac{x^2}{2} - \frac{x^4}{4} + \cdots + (-1)^{r-1}\frac{x^{2r-2}}{2r-2}$:

$$\frac{u_{r+1}}{u_r} = \frac{(-1)^r \frac{x^{2r}}{2r}}{(-1)^{r-1}\frac{x^{2r-2}}{2r-2}} = \frac{-x^2(2r-2)}{2r} = -\left(1-\frac{1}{r}\right)x^2 = \frac{x^2}{r} - x^2$$

$$\lim_{r\to\infty}\left|\frac{u_{r+1}}{u_r}\right| = \left|\frac{x^2}{r} - x^2\right|$$

$$= \left|-x^2\right|, \therefore \text{ converges for } \left|-x^2\right| < 1, \textit{i.e.} \text{ when } -1 < x < 1.$$

1.5. (a)

$f(x) = e^{-x}$ $f^{(1)}(x) = -e^{-x}$ $f^{(2)}(x) = e^{-x}$ $f^{(3)}(x) = -e^{-x}$
$f^{(4)}(x) = e^{-x}$ $f^{(5)}(x) = -e^{-x}$

$f(0) = 1$ $f^{(1)}(0) = -1$ $f^{(2)}(0) = 1$ $f^{(3)}(0) = -1$
$f^{(4)}(0) = 1$ $f^{(5)}(0) = -1$

$$\therefore f(x) = 1 - x + \frac{x^2}{2!} - \frac{x^3}{3!} + \cdots \qquad U_n = \frac{(-1)^{n-1}x^{n-1}}{(n-1)!}.$$

(b)

$f(x) = \cos x$ $f^{(1)}(x) = -\sin x$ $f^{(2)}(x) = -\cos x$ $f^{(3)}(x) = \sin x$
$f^{(4)}(x) = \cos x$ $f^{(5)}(x) = -\sin x$

$f(0) = 1$ $f^{(1)}(0) = 0$ $f^{(2)}(0) = -1$ $f^{(3)}(0) = 0$
$f^{(4)}(0) = 1$ $f^{(5)}(0) = 0$

$$f(x) = 1 - \frac{x^2}{2!} + \frac{x^4}{4!} - \frac{x^6}{6!} + \cdots, \qquad U_n = \frac{(-1)^{n-1}x^{2n-2}}{(2n-2)!}.$$

(c)

$f(x) = (1-x)^{-1}$ $f^{(1)}(x) = (1-x)^{-2}$ $f^{(2)}(x) = 1 \times 2(1-x)^{-3}$
$f^{(3)}(x) = 1 \times 2 \times 3(1-x)^{-4}$

$f(0) = 1$ $f^{(1)}(0) = 1$ $f^{(2)}(0) = 2$ $f^{(3)}(0) = 6$

$$f(x) = 1 + x + \frac{2!x^2}{2!} + \frac{3!x^3}{3!} - \cdots$$
$$\Rightarrow f(x) = 1 + x + x^2 + x^3 + \cdots, \qquad U_n = x^{n-1}.$$

1.6. (a)

$f(x) = (1-x)^{-1}$ $\quad f^{(1)}(x) = (1-x)^{-2}$ $\quad f^{(2)}(x) = 1 \times 2(1-x)^{-3}$ $\quad f^{(3)}(x) = 1 \times 2 \times 3(1-x)^{-4}$

$f(-1) = \frac{1}{2}$ $\quad f^{(1)}(-1) = \frac{1}{4}$ $\quad f^{(2)}(-1) = \frac{2}{8}$ $\quad f^{(3)}(-1) = \frac{6}{16}$

$$f(x) = \frac{1}{2} + \frac{1}{4}(x+1) + \frac{2(x+1)^2}{8 \times 2!} + \frac{6(x+1)^3}{16 \times 3!} + \cdots$$

$$f(x) = \frac{1}{2} + \frac{(x+1)}{4} + \frac{(x+1)^2}{8} + \frac{(x+1)^3}{16} + \cdots, \qquad U_n = \frac{(x+1)^{n-1}}{2^n}.$$

(b)

$f(x) = \sin x$ $\quad f^{(1)}(x) = \cos x$ $\quad f^{(2)}(x) = -\sin x$ $\quad f^{(3)}(x) = -\cos x$ $\quad f^{(4)}(x) = \sin x$ $\quad f^{(5)}(x) = \cos x$

$f(\frac{\pi}{2}) = 1$ $\quad f^{(1)}(\frac{\pi}{2}) = 0$ $\quad f^{(2)}(0) = -1$ $\quad f^{(3)}(0) = 0$ $\quad f^{(4)}(0) = 1$ $\quad f^{(5)}(0) = 0$

$$f(x) = 1 - \frac{1}{2!}(x - \frac{\pi}{2})^2 + \frac{1}{4!}(x - \frac{\pi}{2})^4 - \frac{1}{6!}(x - \frac{\pi}{2})^6 \ldots,$$
$$U_n = \frac{(-1)^{n-1}}{(2n-2)!}(x - \frac{\pi}{2})^{2n-2}.$$

(c)

$f(x) = \ln x$ $\quad f^{(1)}(x) = 1/x$ $\quad f^{(2)}(x) = -1/x^2$ $\quad f^{(3)}(x) = 2/x^3$ $\quad f^{(4)}(x) = -6/x^4$ $\quad f^{(5)}(x) = 24/x^5$

$f(1) = 0$ $\quad f^{(1)}(1) = 1$ $\quad f^{(2)}(1) = -1$ $\quad f^{(3)}(0) = 2$ $\quad f^{(4)}(1) = -6$ $\quad f^{(5)}(1) = 24$

$$f(x) = 0 + (x-1) - \frac{1}{2!}(x-1)^2 + \frac{2}{3!}(x-1)^3 - \frac{6}{4!}(x-1)^4 \ldots$$
$$f(x) = (x-1) - \frac{(x-1)^2}{2} + \frac{(x-1)^3}{3} - \frac{(x-1)^4}{4} \ldots,$$
$$U_n = (-1)^{n-1}\frac{(x-1)^n}{n}.$$

1.7. (a) (i) $E(R) = D_e\left\{1 - e^{-\alpha(R-R_e)}\right\}^2$. Let $u = 1 - e^{-\alpha(R-R_e)} \Rightarrow \frac{du}{dR} = \alpha e^{-\alpha(R-R_e)}$.

$$E(R) = D_e u^2 \Rightarrow \frac{dE(R)}{du} = 2D_e u = 2D_e\left\{1 - e^{-\alpha(R-R_e)}\right\}$$

$$\Rightarrow \frac{dE(R)}{dR} = \frac{dE(R)}{du} \times \frac{du}{dR} = 2D_e\left\{1 - e^{-\alpha(R-R_e)}\right\} \times \alpha e^{-\alpha(R-R_e)} = E^{(1)}R.$$

(ii) $E^{(1)}(R) = \underbrace{2D_e\left\{1 - e^{-\alpha(R-R_e)}\right\}}_{u} \times \underbrace{\alpha e^{-\alpha(R-R_e)}}_{v}$. Using the product rule:

$$E^{(2)}(R) = 2D_e\left\{1 - e^{-\alpha(R-R_e)}\right\} \times -\alpha^2 e^{-\alpha(R-R_e)} + \alpha e^{-\alpha(R-R_e)} \times 2\alpha D_e e^{-\alpha(R-R_e)}$$

$$= -2\alpha^2 D_e e^{-\alpha(R-R_e)} + 2\alpha^2 D_e e^{-2\alpha(R-R_e)} + 2\alpha^2 e^{-2\alpha(R-R_e)}$$

$$\Rightarrow E^{(2)}(R) = 4\alpha^2 D_e e^{-2\alpha(R-R_e)} - 2\alpha^2 D_e e^{-\alpha(R-R_e)}$$

$$\Rightarrow E^{(2)}(R) = 2\alpha^2 D_e\left\{2e^{-2\alpha(R-R_e)} - e^{-\alpha(R-R_e)}\right\}.$$

(b) $E^{(1)}(R_e) = 2D_e\left\{1 - e^{-\alpha(R_e-R_e)}\right\} \times \alpha e^{-\alpha(R_e-R_e)} = 2D_e\{1-1\} \times \alpha = 0$; therefore a maximum or minimum.

$E^{(2)}(R_e) = 2\alpha^2 D_e\left\{2e^{-2\alpha(R_e-R_e)} - e^{-\alpha(R_e-R_e)}\right\} = 2\alpha^2 D_e\{2-1\} = 2\alpha^2 D_e$; positive; therefore a minimum.

(c) $E(R) = \alpha^2 D_e(R - R_e)^2 \Rightarrow$ Let $u = R - R_e \Rightarrow \frac{du}{dR} = 1$.

$$E(R) = \alpha^2 D_e u^2 \Rightarrow \frac{dE(R)}{du} = 2\alpha^2 D_e u = 2\alpha^2 D_e(R - R_e)$$

$$\Rightarrow \frac{dE(R)}{dR} = \frac{dE(R)}{du} \times \frac{du}{dR} = 2\alpha^2 D_e(R - R_e) \times 1 = 2\alpha^2 D_e(R - R_e).$$

Thus $F = -\frac{dE(R)}{dR} = -2\alpha^2 D_e(R - R_e)$.

(d) If $k = 2\alpha^2 D_e$ and $x = R - R_e$, then $F = -kx \Rightarrow$ the expression for F obtained in (c) has the same form as the restoring force acting on a simple harmonic oscillator.

1.8. (a) $\sinh x = \frac{1}{2}(e^x - e^{-x})$

$$= \frac{1}{2}\left\{\left(1 + x + \frac{x^2}{2!} + \frac{x^3}{3!} + \cdots + \frac{x^{n-1}}{(n-1)!} + \cdots\right)\right.$$

$$\left. - \left(1 - x + \frac{x^2}{2!} - \frac{x^3}{3!} + \frac{x^4}{4!} - \cdots (-1)^{n-1}\frac{x^{n-1}}{(n-1)!} + \cdots\right)\right\}$$

$$= \frac{1}{2}\left\{2x + \frac{2x^3}{3!} + \cdots + \frac{2x^{2n-1}}{(2n-1)!} + \cdots\right\}$$

$$= x + \frac{x^3}{3!} + \frac{x^5}{5!} + \cdots + \frac{x^{2n-1}}{(2n-1)!} + \cdots$$

(b) $f(x) = \dfrac{e^{-x}}{(1-x)} = e^{-x} \times \dfrac{1}{(1-x)}$

$$= \left\{1 - x + \frac{x^2}{2!} - \frac{x^3}{3!} + \frac{x^4}{4!} - \cdots (-1)^n \frac{x^n}{n!} + \cdots\right\}$$

$$\times\left\{1 + x + x^2 + x^3 + \cdots x^n + \cdots\right\}$$

$$= 1 + \frac{x^2}{2!} + \left\{\frac{x^3}{2!} - \frac{x^3}{3!}\right\} + \left\{\frac{x^4}{2!} - \frac{x^4}{3!} + \frac{x^4}{4!}\right\} + \cdots$$

$$= 1 + \frac{x^2}{2} + \frac{x^3}{3} + \cdots$$

The Maclaurin power series expansion of e^{-x} converges for all x, and of $\frac{1}{(1-x)}$ converges for $-1 < x < 1$, so the interval of convergence of the Maclaurin series of $f(x) = \frac{e^{-x}}{(1-x)}$ is $-1 < x < 1$.

1.9. (a) $$e^X = 1 + X + \frac{X^2}{2!} + \frac{X^3}{3!} + \frac{X^4}{4!} + \cdots \frac{X^{n-1}}{(n-1)!} + \cdots;$$

if $X = ax$, then:

$$e^{ax} = 1 + ax + \frac{(ax)^2}{2!} + \frac{(ax)^3}{3!} + \frac{(ax)^4}{4!} + \cdots \frac{(ax)^{n-1}}{(n-1)!} + \cdots$$

(b) (i) $\sin 2x = 2\sin x\cos x$

$$= 2\left\{x - \frac{x^3}{3!} + \frac{x^5}{5!} - \cdots\right\} \times \left\{1 - \frac{x^2}{2!} + \frac{x^4}{4!} - \cdots\right\}$$

$$\sin 2x = 2x - \frac{8x^3}{6} + \frac{32x^5}{120} - \cdots = 2x - \frac{8x^3}{3!} + \frac{32x^5}{5!} - \cdots$$

(ii) $\sin X = X - \dfrac{X^3}{3!} + \dfrac{X^5}{5!} - \cdots$; if $X = 2x$, then:

$$\sin 2x = 2x - \frac{(2x)^3}{3!} + \frac{(2x)^5}{5!} - \cdots$$

$$\Rightarrow \sin 2x = 2x - \frac{8x^3}{3!} + \frac{32x^5}{5!} - \cdots$$

1.10. $C_V = 3R\left(\frac{h\nu}{kT}\right)^2\left\{\frac{e^{h\nu/2kT}}{e^{h\nu/2kT}-1}\right\}^2$

$$e^{h\nu/2kT} = 1 + \frac{h\nu}{2k}\frac{1}{T} + \frac{\left(\frac{h\nu}{2k}\right)^2\frac{1}{T^2}}{2!} + \cdots \approx 1, \text{ for large } T.$$

$$e^{h\nu/kT} - 1 = 1 + \frac{h\nu}{k}\frac{1}{T} + \frac{\left(\frac{h\nu}{k}\right)^2\frac{1}{T^2}}{2!} + \cdots - 1$$

$$= \frac{h\nu}{k}\frac{1}{T} + \frac{\left(\frac{h\nu}{k}\right)^2\frac{1}{T^2}}{2!} + \cdots \approx \frac{h\nu}{k}\frac{1}{T}, \text{ for large } T.$$

Therefore, $\lim_{\text{large } T} C_V = 3R\left(\frac{h\nu}{kT}\right)^2\left\{\frac{1}{h\nu/kT}\right\}^2 = 3R\left(\frac{h\nu}{kT}\right)^2\left\{\frac{kT}{h\nu}\right\}^2 = 3R.$

1.11. (a) $\sin\frac{n\pi x}{L} \approx \frac{n\pi x}{L}$ for small x. At $x = 0$, $\sin\frac{n\pi x}{L} = 0 \Rightarrow \psi = 0$.

(b) At $x = L$, $\psi = \sqrt{\frac{2}{L}}\sin n\pi = 0$, when $n = 1, 2, 3, \ldots$

(c) When x is small, $\psi = \sqrt{\frac{2}{L}}\frac{n\pi}{L}x$.

Chapter 2

2.1. (a)

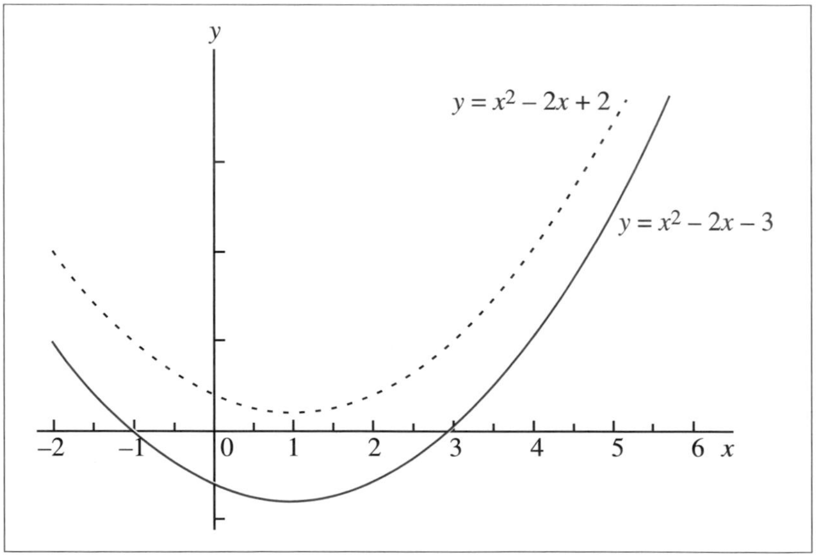

(b) (i) $x^2 - 2x - 3 = 0 \Rightarrow x = \dfrac{2 \pm \sqrt{4 - -(4 \times 1 \times 3)}}{2} = 1 \pm \dfrac{\sqrt{16}}{2}$

$= 1 \pm 2 = 3,\ -1.$

(ii) $x^2 - 2x + 2 = 0 \Rightarrow x = \dfrac{2 \pm \sqrt{4 - (4 \times 1 \times 2)}}{2} = 1 \pm \dfrac{\sqrt{-4}}{2}$

$= 1 \pm \dfrac{\sqrt{4}\sqrt{-1}}{2} = 1 \pm \dfrac{2\mathrm{i}}{2} = 1 \pm \mathrm{i}.$

Part (i) has two real roots corresponding to where the curve cuts the x-axis. For part (ii), there are no real roots and so the curve does not cut the x axis.

2.2. (a) $\mathrm{i}^3 = \mathrm{i} \times \mathrm{i}^2 = \mathrm{i} \times -1 = -\mathrm{i}.$
(b) $\mathrm{i}^4 = \mathrm{i} \times \mathrm{i}^3 = \mathrm{i} \times -\mathrm{i} = 1.$
(c) $\mathrm{i}^5 = \mathrm{i} \times \mathrm{i}^4 = \mathrm{i} \times 1 = \mathrm{i}.$

2.3. (a) $z_1 + z_2 = (2 + 3\mathrm{i}) + (-1 + \mathrm{i}) = 1 + 4\mathrm{i}.$

$$\therefore\ z_1 + z_2 - 2z_3 = (1 + 4\mathrm{i}) - 2(3 - 2\mathrm{i}) = -5 + 8\mathrm{i}.$$

(b) $z_1 z_2 = (2 + 3\mathrm{i})(-1 + \mathrm{i}) = -2 + 2\mathrm{i} - 3\mathrm{i} + 3\mathrm{i}^2 = -2 - \mathrm{i} - 3 = -5 - \mathrm{i}.$

$$z_3^2 = (3 - 2\mathrm{i})(3 - 2\mathrm{i}) = 9 - 6\mathrm{i} - 6\mathrm{i} + 4\mathrm{i}^2 = 9 - 12\mathrm{i} - 4 = 5 - 12\mathrm{i}.$$

$$\therefore\ z_1 z_2 + z_3^2 = (-5 - \mathrm{i}) + (5 - 12\mathrm{i}) = -13\mathrm{i}.$$

2.4. (a) $z = (-1 - 2\mathrm{i}) + (2 + 7\mathrm{i}) = 1 + 5\mathrm{i};\ z^* = 1 - 5\mathrm{i}.$
(b) $z = (3 - \mathrm{i}) - (4 - 2\mathrm{i}) = -1 + \mathrm{i};\ z^* = -1 - \mathrm{i}.$
(c) $z = \mathrm{i}(1 + 3\mathrm{i}) = \mathrm{i} - 3 = -3 + \mathrm{i};\ z^* = -3 - \mathrm{i}.$
(d) $z = (1 + 3\mathrm{i})(3 + 2\mathrm{i}) = 3 + 2\mathrm{i} + 9\mathrm{i} - 6 = -3 + 11\mathrm{i};\ z^* = -3 - 11\mathrm{i}.$

2.5. (a) $\dfrac{1}{\mathrm{i}} = \dfrac{1}{\mathrm{i}} \times \dfrac{-\mathrm{i}}{-\mathrm{i}} = \dfrac{-\mathrm{i}}{1} = -\mathrm{i}.$

(b) $\dfrac{1 - \mathrm{i}}{2 - \mathrm{i}} = \dfrac{1 - \mathrm{i}}{2 - \mathrm{i}} \times \dfrac{2 + \mathrm{i}}{2 + \mathrm{i}} = \dfrac{2 + \mathrm{i} - 2\mathrm{i} + 1}{4 + 2\mathrm{i} - 2\mathrm{i} + 1} = \dfrac{3 - \mathrm{i}}{5} = \dfrac{3}{5} - \dfrac{\mathrm{i}}{5}.$

(c) $\dfrac{\mathrm{i}(2 + \mathrm{i})}{(1 - 2\mathrm{i})(2 - \mathrm{i})} = \dfrac{-1 + 2\mathrm{i}}{2 - \mathrm{i} - 4\mathrm{i} - 2} = \dfrac{-1 + 2\mathrm{i}}{-5\mathrm{i}} = \dfrac{-1 + 2\mathrm{i}}{-5\mathrm{i}} \times \dfrac{5\mathrm{i}}{5\mathrm{i}}$

$= \dfrac{-5\mathrm{i} - 10}{25} = \dfrac{-2}{5} - \dfrac{\mathrm{i}}{5}.$

2.6.

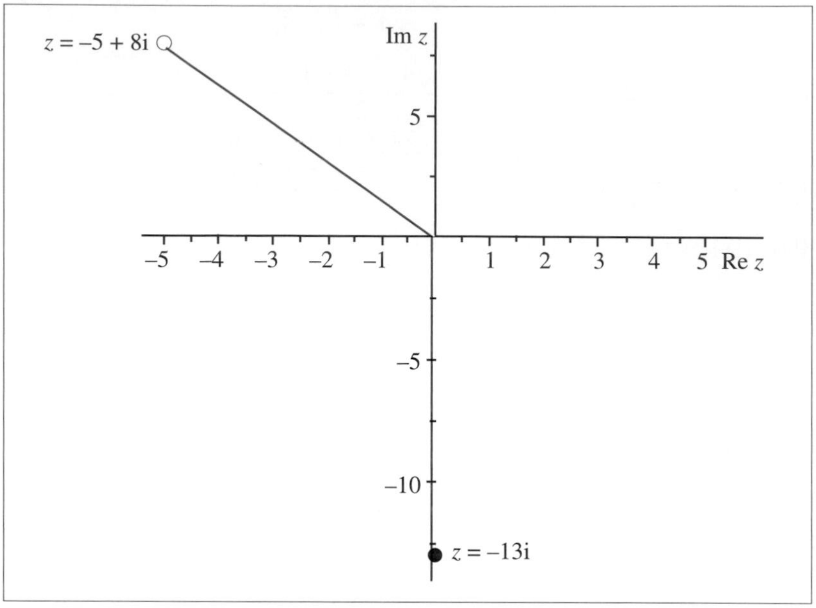

2.7. (a) $z = -1 - 2\text{i}$; $r = \sqrt{1^2 + 2^2} = \sqrt{5}$; $\theta = \tan^{-1}(-2/-1) = \tan^{-1}2 = 63.43°, -116.56°$. z lies in the 3rd quadrant and so $\theta = -116.56°$.
(b) $z = 2\text{i}$; $r = \sqrt{2^2} = 2$; $\theta = \tan^{-1}(2/0)$; undefined at $\theta = 90°$, $-90°$. In this instance, $\theta = 90°$.

2.8. (a) $(\text{i}\theta)^2 = -\theta^2$; (b) $(\text{i}\theta)^3 = -\text{i}\theta^3$; (c) $(\text{i}\theta)^4 = \theta^4$; (d) $(\text{i}\theta)^5 = \text{i}\theta^5$.

$$z = r\left\{1 + \text{i}\theta + \frac{(\text{i}\theta)^2}{2!} + \frac{(\text{i}\theta)^3}{3!} + \frac{(\text{i}\theta)^4}{4!} + \frac{(\text{i}\theta)^5}{5!} + \ldots\right\}$$
$$= r\left\{1 + \text{i}\theta - \frac{\theta^2}{2!} - \frac{\text{i}\theta^3}{3!} + \frac{\theta^4}{4!} + \frac{\text{i}\theta^5}{5!} + \ldots\right\}\text{: same as equation (2.16).}$$

2.9. (a) $z_1 z_2 = r_1 \text{e}^{\text{i}\theta_1} \times r_2 \text{e}^{\text{i}\theta_2} = r_1 r_1 \text{e}^{\text{i}(\theta_1 + \theta_2)}$; modulus $= r_1 r_2$, argument $= \theta_1 + \theta_2$.

(b) $\dfrac{z_1}{z_2} = \dfrac{r_1 \text{e}^{\text{i}\theta_1}}{r_2 \text{e}^{\text{i}\theta_2}} = \dfrac{r_1}{r_2}\text{e}^{\text{i}\theta_1} \times \text{e}^{-\text{i}\theta_2} = \dfrac{r_1}{r_2}\text{e}^{\text{i}(\theta_1 - \theta_2)}$; modulus $= \dfrac{r_1}{r_2}$, argument $= \theta_1 - \theta_2$.

(c) $\frac{z_1^2}{z_2^4} = \frac{r_1^2 e^{2i\theta_1}}{r_2^4 e^{4i\theta_2}} = \frac{r_1^2}{r_2^4} e^{2i\theta_1} e^{-4i\theta_2} = \frac{r_1^2}{r_2^4} e^{i(2\theta_1 - 4\theta_2)}$; modulus $= \frac{z_1^2}{z_2^4}$, argument $= 2\theta_1 - 4\theta_2$.

2.10. $z = -1 - i$, $r = \sqrt{1^2 + 1^2} = \sqrt{2}$; $\theta = \tan^{-1}(-1/-1) = \tan^{-1} 1 = \pi/4, -3\pi/4$, but in 3rd quadrant and so $\theta = -3\pi/4$.

Thus, $z = \sqrt{2} e^{-\frac{3\pi}{4} i} \Rightarrow z^2 = 2e^{-\frac{6\pi}{4} i} = 2e^{-\frac{3\pi}{2} i}$; modulus $= 2$, argument $= -3\pi/2 = \pi/2$.

$z^{-4} = \frac{1}{4} e^{3\pi i}$; modulus $= 1/4$, argument $= 3\pi = \pm\pi$, but from the definition of the argument, $\theta = \pi$ is the only acceptable result.

2.11. (a) $(\cos\theta + i\sin\theta)^{-1} = \cos -\theta + i\sin -\theta = \cos\theta - i\sin\theta$.
(b) $(\cos\theta + i\sin\theta)^{1/2} = \cos\theta/2 + i\sin\theta/2$.
(c) $z^n = \underbrace{r^n \cos n\theta}_{\text{real}} + \underbrace{i\, r^n \sin n\theta}_{\text{imaginary}}$

(d) $z = -1 - i$; $r = \sqrt{2}$; $\theta = -3\pi/4 \Rightarrow z = \sqrt{2}\{\cos(3\pi/4) - i\sin(3\pi/4)\}$.

$$z^3 = \sqrt{2}^3 \{\cos(9\pi/4) - i\sin(9\pi/4)\} = \sqrt{2}^3 \sqrt{2}^{-1} - i\sqrt{2}^3 \sqrt{2}^{-1} = 2 - 2i.$$

$$\begin{aligned} z^{-2} &= \sqrt{2}^{-2} \{\cos(-3\pi/2) - i\sin(-3\pi/2)\} \\ &= \frac{1}{2}\{\cos(3\pi/2) + i\sin(3\pi/2)\} = \frac{1}{2} \times -i = -\frac{i}{2}. \end{aligned}$$

2.12. $e^{i2m\pi} = \cos 2m\pi + i\sin 2m\pi = 1$; $e^{i(\theta + 2m\pi)} = e^{i\theta} e^{i2m\pi} = e^{i\theta}$.

2.13. (a) $e^{-i\theta} = \cos(-\theta) + i\sin(-\theta) = \cos\theta - i\sin\theta$.
(b) $e^{-i\theta} = \cos\theta - i\sin\theta$ and $e^{i\theta} = \cos\theta + i\sin\theta$.
(i) Adding the two expressions yields:

$$e^{i\theta} + e^{-i\theta} = 2\cos\theta + i\sin\theta - i\sin\theta = 2\cos\theta \Rightarrow \cos\theta = \frac{1}{2}\left(e^{i\theta} + e^{-i\theta}\right).$$

(ii) Similarly, subtracting the two expressions yields:

$$\begin{aligned} &e^{i\theta} - e^{-i\theta} = \cos\theta - \cos\theta + i\sin\theta + i\sin\theta = 2i\sin\theta \\ &\Rightarrow \sin\theta = \frac{1}{2i}\left(e^{i\theta} - e^{-i\theta}\right). \end{aligned}$$

2.14. $y = A\cos kt + B\sin kt$

$$\therefore y = A \times \frac{1}{2}(\mathrm{e}^{\mathrm{i}kt} + \mathrm{e}^{-\mathrm{i}kt}) + B \times \frac{1}{2\mathrm{i}}(\mathrm{e}^{\mathrm{i}kt} - \mathrm{e}^{-\mathrm{i}kt})$$
$$= \frac{A}{2}\mathrm{e}^{\mathrm{i}kt} + \frac{A}{2}\mathrm{e}^{-\mathrm{i}kt} + \frac{B}{2\mathrm{i}}\mathrm{e}^{\mathrm{i}kt} - \frac{B}{2\mathrm{i}}\mathrm{e}^{-\mathrm{i}kt}$$
$$= \left(\frac{A}{2} + \frac{B}{2\mathrm{i}}\right)\mathrm{e}^{\mathrm{i}kt} + \left(\frac{A}{2} - \frac{B}{2\mathrm{i}}\right)e^{-\mathrm{i}kt}.$$

2.15. (a) $\psi_1 = \underbrace{N_1\mathrm{e}^{-r/2a_0}r\sin\theta}_{\text{real}}\ \underbrace{\mathrm{e}^{\mathrm{i}\phi}}_{\text{imaginary}}$.

$\psi_0 = N_2\mathrm{e}^{-r/2a_0}r\cos\theta$; real, no imaginary part.

$\psi_{-1} = \underbrace{N_1\mathrm{e}^{-r/2a_0}r\sin\theta}_{\text{real}}\ \underbrace{\mathrm{e}^{-\mathrm{i}\phi}}_{\text{imaginary}}$.

(b) (i) $\psi_1 + \psi_{-1} = N_1\mathrm{e}^{-r/2a_0}r\sin\theta\mathrm{e}^{\mathrm{i}\phi} + N_1\mathrm{e}^{-r/2a_0}r\sin\theta\mathrm{e}^{-\mathrm{i}\phi}$ $= N_1\mathrm{e}^{-r/2a_0}r\ \sin\theta(\mathrm{e}^{\mathrm{i}\phi} + \mathrm{e}^{-\mathrm{i}\phi})$, but $\mathrm{e}^{\mathrm{i}\phi} + \mathrm{e}^{-\mathrm{i}\phi} = 2\cos\phi$ and so:

$$\frac{1}{\sqrt{2}}(\psi_1 + \psi_{-1}) = \frac{2}{\sqrt{2}}N_1\mathrm{e}^{-r/2a_0}r\sin\theta \times \cos\phi$$
$$= \sqrt{2}N_1\mathrm{e}^{-r/2a_0}r\sin\theta\cos\phi.$$

(ii) $\psi_1 - \psi_{-1} = N_1\mathrm{e}^{-r/2a_0}r\sin\theta\mathrm{e}^{\mathrm{i}\phi} - N_1\mathrm{e}^{-r/2a_0}r\sin\theta\mathrm{e}^{-\mathrm{i}\phi}$ $= N_1\mathrm{e}^{-r/2a_0}\ r\sin\theta(\mathrm{e}^{\mathrm{i}\phi} - \mathrm{e}^{-\mathrm{i}\phi})$, but $\mathrm{e}^{\mathrm{i}\phi} - \mathrm{e}^{-\mathrm{i}\phi} = 2\mathrm{i}\sin\phi$ and so:

$$\frac{-\mathrm{i}}{\sqrt{2}}(\psi_1 - \psi_{-1}) = \frac{-\mathrm{i}}{\sqrt{2}}N_1\mathrm{e}^{-r/2a_0}r\sin\theta \times 2\mathrm{i}\sin\phi$$
$$= \sqrt{2}N_1\mathrm{e}^{-r/2a_0}r\sin\theta\sin\phi.$$

(c) $\psi_0 = N_2\mathrm{e}^{-r/2a_0}r\cos\theta$; $z = r\cos\theta \Rightarrow \psi_0 = N_2\mathrm{e}^{-r/2a_0}z$. Thus we can relabel ψ_0 as ψ_z.

$$\frac{1}{\sqrt{2}}(\psi_1 + \psi_{-1}) = \sqrt{2}N_1\mathrm{e}^{-r/2a_0}r\sin\theta\cos\phi; \qquad x = r\sin\theta\cos\phi$$
$$\Rightarrow \psi_x = \sqrt{2}N_1\mathrm{e}^{-r/2a_0}x.$$
$$\frac{-\mathrm{i}}{\sqrt{2}}(\psi_1 - \psi_{-1}) = \sqrt{2}N_1\mathrm{e}^{-r/2a_0}r\sin\theta\sin\phi; \qquad y = r\sin\theta\sin\phi$$
$$\Rightarrow \psi_y = \sqrt{2}N_1\mathrm{e}^{-r/2a_0}y.$$

2.16. (a) $F(hkl) = \sum_{j}^{\text{cell}} f_j e^{2\pi i[hx_j + ky_j + lz_j]}$

$$\Rightarrow F(hkl) = f_{Na} e^{2\pi i[h0 + k0 + l0]} + f_{Na} e^{2\pi i[h\frac{1}{2} + k\frac{1}{2} + l\frac{1}{2}]} = f_{Na} + f_{Na} e^{\pi i[h+k+l]}.$$

Euler's formula: $e^{i\theta} = \cos\theta + i\sin\theta$

$$\Rightarrow F(hkl) = f_{Na} + f_{Na} e^{\pi i[h+k+l]}$$
$$= f_{Na} + f_{Na}\{\cos(h+k+l)\pi + i\sin(h+k+l)\pi\}.$$

(b) $$F(hkl) = f_{Na} + f_{Na}\{\underbrace{\cos(h+k+l)\pi}_{\substack{=1 \text{ when } h+k+l \text{ even} \\ =-1 \text{ when } h+k+l \text{ odd}}} + \underbrace{i\sin(h+k+l)\pi}_{=0 \text{ for } h+k+l \text{ even, odd}}\}$$

$$\therefore F(hkl) = 2f_{Na}, \text{ for } n \text{ even}; = 0 \text{ for } n \text{ odd}.$$

2.17. For the complex number i, $r = 1$ and $\theta = \frac{\pi}{2}$; hence:

$$i^{1/3} = e^{i(\pi/2 + 2m\pi) \times 1/3} = e^{i(\pi/6 + 2m\pi/3)}$$
$$= \cos\left(\frac{\pi}{6} + \frac{2m\pi}{3}\right) + i\sin\left(\frac{\pi}{6} + \frac{2m\pi}{3}\right)$$

For $m = 1$:

$$i^{1/3} = \cos\left(\frac{5\pi}{6}\right) + i\sin\left(\frac{5\pi}{6}\right) = -\frac{\sqrt{3}}{2} + \frac{1}{2}i.$$

For $m = 2$:

$$i^{1/3} = \cos\left(\frac{3\pi}{2}\right) + i\sin\left(\frac{3\pi}{2}\right) = -i;$$

For $m = 3$:

$$i^{1/3} = \cos\left(\frac{13\pi}{6}\right) + i\sin\left(\frac{13\pi}{6}\right) = \frac{\sqrt{3}}{2} + \frac{1}{2}i.$$

For $m = 4$:

$$i^{1/3} = \cos\left(\frac{17\pi}{6}\right) + i\sin\left(\frac{17\pi}{6}\right) = -\frac{\sqrt{3}}{2} + \frac{1}{2}i,$$

and so on. We see that taking $m \geqslant 4$ merely replicates the roots already found, and so the three cube roots of i are $-\frac{\sqrt{3}}{2} + \frac{1}{2}i$, $-i$ and $\frac{\sqrt{3}}{2} + \frac{1}{2}i$.

Chapter 3

3.1. $2x + y = 5$ and $\frac{1}{2}x + 8y = 9$:

$$x = \frac{\begin{vmatrix} 5 & 9 \\ 1 & 8 \end{vmatrix}}{\begin{vmatrix} 2 & 1 \\ \frac{1}{2} & 8 \end{vmatrix}} = \frac{40 - 9}{16 - \frac{1}{2}} = \frac{31}{15\frac{1}{2}} = 2;$$

$$y = \frac{\begin{vmatrix} 9 & 5 \\ \frac{1}{2} & 2 \end{vmatrix}}{\begin{vmatrix} 2 & 1 \\ \frac{1}{2} & 8 \end{vmatrix}} = \frac{18 - 2\frac{1}{2}}{16 - \frac{1}{2}} = \frac{15\frac{1}{2}}{15\frac{1}{2}} = 1.$$

3.2. (a) $a_{11} \equiv 1,\ a_{12} \equiv -\dfrac{1}{RT_1},\ b_1 \equiv \ln k_1;\quad a_{21} \equiv 1,\ a_{22} \equiv -\dfrac{1}{RT_2},$ $b_2 \equiv \ln k_2;\ x = \ln A,\ y = E_a$.

(b) $a_{11} \equiv 100,\ a_{12} \equiv 1,\ b_1 \equiv 212;\ a_{21} \equiv 0,\ a_{22} \equiv 1,\ b_2 \equiv 32;$

$$a = \frac{\begin{vmatrix} 212 & 32 \\ 1 & 1 \end{vmatrix}}{\begin{vmatrix} 100 & 1 \\ 0 & 1 \end{vmatrix}} = \frac{212 - 32}{100} = \frac{180}{100} = \frac{9}{5} \text{ and } b = \frac{\begin{vmatrix} 32 & 212 \\ 0 & 100 \end{vmatrix}}{\begin{vmatrix} 100 & 1 \\ 0 & 1 \end{vmatrix}} = \frac{3200}{100} = 32$$

and so the formula relating T to t is $T = \frac{9}{5}t + 32$.

3.3. (a)
$$\begin{vmatrix} 1 & -1 & 2 \\ 0 & 3 & 0 \\ 2 & -2 & -2 \end{vmatrix} = 1\begin{vmatrix} 0 & 0 \\ 2 & -2 \end{vmatrix} + 3\begin{vmatrix} 1 & 2 \\ 2 & -2 \end{vmatrix} + 2\begin{vmatrix} 1 & 2 \\ 0 & 0 \end{vmatrix}$$
$$= 0 + 3(-2 - 4) + 0 = -18.$$

(b)
$$\begin{vmatrix} 1 & -1 & 2 \\ 0 & 3 & 0 \\ 2 & -2 & -2 \end{vmatrix} = 0\begin{vmatrix} -1 & 2 \\ -2 & -2 \end{vmatrix} + 3\begin{vmatrix} 1 & 2 \\ 2 & -2 \end{vmatrix} + 0\begin{vmatrix} 1 & -1 \\ 2 & -2 \end{vmatrix}$$
$$= 0 + 3(-2 - 4) + 0 = -18.$$

3.4. (a)
$$\begin{vmatrix} 1 & 0 & -2 \\ 2 & 8 & 4 \\ 3 & 2 & 2 \end{vmatrix} \quad A_{33} = (-1)^6\begin{vmatrix} 1 & 0 \\ 2 & 8 \end{vmatrix} = 8;$$

$$A_{22} = (-1)^4\begin{vmatrix} 1 & -2 \\ 3 & 2 \end{vmatrix} = 2 - -6 = 8;$$

$$A_{32} = (-1)^5\begin{vmatrix} 1 & -2 \\ 2 & 4 \end{vmatrix} = -1(4 - -4) = -8;$$

$$A_{23} = (-1)^5\begin{vmatrix} 1 & 0 \\ 3 & 2 \end{vmatrix} = -1 \times 2 = -2.$$

(b) $\begin{vmatrix} \cos\theta & -\sin\theta & 0 \\ \sin\theta & \cos\theta & 0 \\ 0 & 0 & 1 \end{vmatrix}$:

$$A_{12} = (-1)^3 \begin{vmatrix} \sin\theta & 0 \\ 0 & 1 \end{vmatrix} = -\sin\theta;$$

$$A_{21} = (-1)^3 \begin{vmatrix} -\sin\theta & 0 \\ 0 & 1 \end{vmatrix} = -1 \times -\sin\theta = \sin\theta.$$

3.5. (a)

(i) $\begin{vmatrix} 1 & 2 & 3 \\ 0 & 8 & 2 \\ -2 & 4 & 2 \end{vmatrix} = 0\begin{vmatrix} 2 & 3 \\ 4 & 2 \end{vmatrix} + 8\begin{vmatrix} 1 & 3 \\ -2 & 2 \end{vmatrix} - 2\begin{vmatrix} 1 & 2 \\ -2 & 4 \end{vmatrix}$

$= 0 + 8(2 - -6) - 2(4 - -4) = 64 - 16 = 48.$

(ii) $\begin{vmatrix} 1 & 2 & 3 \\ 0 & 8 & 2 \\ -2 & 4 & 2 \end{vmatrix} \Rightarrow$ Subtract twice col 1 from col 2 $\Rightarrow \begin{vmatrix} 1 & 0 & 3 \\ 0 & 8 & 2 \\ -2 & 8 & 2 \end{vmatrix}$

and then 3 times col 1 from col 3 $\Rightarrow \begin{vmatrix} 1 & 0 & 0 \\ 0 & 8 & 2 \\ -2 & 8 & 8 \end{vmatrix} = 1\begin{vmatrix} 8 & 2 \\ 8 & 8 \end{vmatrix}$

$= 64 - 16 = 48.$

(iii) Starting from $\begin{vmatrix} 1 & 0 & 0 \\ 0 & 8 & 2 \\ -2 & 8 & 8 \end{vmatrix}$, subtract 1/4 col 2 from col 3

$\rightarrow \begin{vmatrix} 1 & 0 & 0 \\ 0 & 8 & 0 \\ -2 & 8 & 6 \end{vmatrix} = 1 \times 8 \times 6 = 48.$

(b) (i) $\begin{vmatrix} 1 & 0 & -2 \\ 2 & 8 & 4 \\ 3 & 2 & 2 \end{vmatrix} = -2\begin{vmatrix} 0 & -2 \\ 2 & 2 \end{vmatrix} + 8\begin{vmatrix} 1 & -2 \\ 3 & 2 \end{vmatrix} - 4\begin{vmatrix} 1 & 0 \\ 3 & 2 \end{vmatrix} =$

$-2(0 - -4) + 8(2 - -6) - 4(2 - 0) = -8 + 64 - 8 = 48.$

(ii) $\begin{vmatrix} 1 & 0 & -2 \\ 2 & 8 & 4 \\ 3 & 2 & 2 \end{vmatrix} \Rightarrow$ Add twice col 1 to col 3 $\Rightarrow \begin{vmatrix} 1 & 0 & 0 \\ 2 & 8 & 8 \\ 3 & 2 & 8 \end{vmatrix} =$

$1\begin{vmatrix} 8 & 8 \\ 2 & 8 \end{vmatrix} = 64 - 16 = 48.$

(iii) Starting from $\begin{vmatrix} 1 & 0 & 0 \\ 2 & 8 & 8 \\ 3 & 2 & 8 \end{vmatrix}$, subtract col 2 from col 3$\Rightarrow$

$\begin{vmatrix} 1 & 0 & 0 \\ 2 & 8 & 0 \\ 3 & 2 & 6 \end{vmatrix} = 1\begin{vmatrix} 8 & 0 \\ 2 & 6 \end{vmatrix} = 1 \times 8 \times 6 = 48.$

3.6. $\begin{vmatrix} \alpha-\varepsilon & \beta & 0 \\ \beta & \alpha-\varepsilon & \beta \\ 0 & \beta & \alpha-\varepsilon \end{vmatrix} = 0$:

(a) $\begin{vmatrix} \frac{\alpha-\varepsilon}{\beta} & 1 & 0 \\ 1 & \frac{\alpha-\varepsilon}{\beta} & 1 \\ 0 & 1 & \frac{\alpha-\varepsilon}{\beta} \end{vmatrix}\beta^3 = 0 \Rightarrow \begin{vmatrix} \frac{\alpha-\varepsilon}{\beta} & 1 & 0 \\ 1 & \frac{\alpha-\varepsilon}{\beta} & 1 \\ 0 & 1 & \frac{\alpha-\varepsilon}{\beta} \end{vmatrix} = 0$

(b) $\begin{vmatrix} x & 1 & 0 \\ 1 & x & 1 \\ 0 & 1 & x \end{vmatrix} = x\begin{vmatrix} x & 1 \\ 1 & x \end{vmatrix} - 1\begin{vmatrix} 1 & 1 \\ 0 & x \end{vmatrix} = x(x^2-1) - (x-0) = x^3 - x - x = x^3 - 2x \Rightarrow \begin{vmatrix} x & 1 & 0 \\ 1 & x & 1 \\ 0 & 1 & x \end{vmatrix} = x^3 - 2x = 0.$

(c) $x^3 - 2x = 0 \Rightarrow x(x^2-2) = 0$, when $x = 0, \ \pm\sqrt{2}$.

(d) $x = (\alpha-\varepsilon)/\beta \Rightarrow \alpha - \beta x = \varepsilon \Rightarrow \varepsilon = \alpha, \ \alpha \mp \sqrt{2}\beta$.

Chapter 4

4.1. $\mathbf{b} = \begin{pmatrix} 1 & 1 & 1 \\ 2 & -2 & 2 \end{pmatrix}$, $\mathbf{c} = \begin{pmatrix} 3 & -1 \\ 1 & -3 \end{pmatrix}$, $\mathbf{d} = \begin{pmatrix} 1 \\ 0 \end{pmatrix}$ and $\mathbf{e} = \begin{pmatrix} 0 & -\mathrm{i} & 1 & \mathrm{i} \end{pmatrix}$:

(a) **b**, rectangular; **c**, square; **d**, column; **e**, row.

(b) $b_{11} = 1, \ b_{12} = 1, \ b_{13} = 1, \ b_{21} = 2, \ b_{22} = -2, \ b_{23} = 2.$
$c_{11} = 3, \ c_{12} = -1, \ c_{21} = 1, \ c_{22} = -3.$
$d_{11} = 1, \ d_{21} = 0.$
$e_{11} = 0, \ e_{12} = -\mathrm{i}, \ e_{13} = 1, \ e_{14} = \mathrm{i}.$

(c) **b**, 2×3; **c**, 2×2; **d**, 2×1; **e**, 1×4.

4.2. (a) $2\mathbf{B} = 2\begin{pmatrix} 4 & 5 \\ 1 & 6 \\ -4 & 3 \end{pmatrix} = \begin{pmatrix} 8 & 10 \\ 2 & 12 \\ -8 & 6 \end{pmatrix}$;

(b) $2\mathbf{C} = 2\begin{pmatrix} 2 & \frac{5}{2} \\ \frac{1}{2} & 3 \\ -2 & \frac{3}{2} \end{pmatrix} = \begin{pmatrix} 4 & 5 \\ 1 & 6 \\ -4 & 3 \end{pmatrix}$.

4.3. (a) $\mathbf{A} + \mathbf{B} = \begin{pmatrix} 1 & \mathrm{i} \\ -\mathrm{i} & 1 \end{pmatrix} + \begin{pmatrix} 1 & -\mathrm{i} \\ \mathrm{i} & 1 \end{pmatrix} = \begin{pmatrix} 2 & 0 \\ 0 & 2 \end{pmatrix} = 2\mathbf{D}$.

(b) $\mathbf{A}-\mathbf{B}=\begin{pmatrix}1 & \mathrm{i}\\ -\mathrm{i} & 1\end{pmatrix}-\begin{pmatrix}1 & -\mathrm{i}\\ \mathrm{i} & 1\end{pmatrix}=\begin{pmatrix}0 & 2\mathrm{i}\\ -2\mathrm{i} & 0\end{pmatrix}=2\mathrm{i}\mathbf{C}.$

(c) $\mathbf{R}+\mathbf{S}=\begin{pmatrix}\cos\theta & \sin\theta\\ -\sin\theta & \cos\theta\end{pmatrix}+\begin{pmatrix}\cos\theta & -\sin\theta\\ \sin\theta & \cos\theta\end{pmatrix}=\begin{pmatrix}2\cos\theta & 0\\ 0 & 2\cos\theta\end{pmatrix}$
$=2\cos\theta\mathbf{D}.$

(d) $\mathbf{R}-\mathbf{S}=\begin{pmatrix}\cos\theta & \sin\theta\\ -\sin\theta & \cos\theta\end{pmatrix}-\begin{pmatrix}\cos\theta & -\sin\theta\\ \sin\theta & \cos\theta\end{pmatrix}=\begin{pmatrix}0 & 2\sin\theta\\ -2\sin\theta & 0\end{pmatrix}$
$=2\sin\theta\mathbf{C}.$

4.4. $\mathbf{AB}=\begin{pmatrix}1 & 2\\ 2 & 1\end{pmatrix}\begin{pmatrix}1 & -1\\ -1 & 2\end{pmatrix}=\begin{pmatrix}-1 & 3\\ 1 & 0\end{pmatrix}: 2\times 2.$

$\mathbf{BA}=\begin{pmatrix}1 & -1\\ -1 & 2\end{pmatrix}\begin{pmatrix}1 & 2\\ 2 & 1\end{pmatrix}=\begin{pmatrix}-1 & 1\\ 3 & 0\end{pmatrix}: 2\times 2.$

$\mathbf{AC}=\begin{pmatrix}1 & 2\\ 2 & 1\end{pmatrix}\begin{pmatrix}-1 & 1\\ -1 & 1\end{pmatrix}=\begin{pmatrix}-3 & 3\\ -3 & 3\end{pmatrix}: 2\times 2.$

$\mathbf{BC}=\begin{pmatrix}1 & -1\\ -1 & 2\end{pmatrix}\begin{pmatrix}-1 & 1\\ -1 & 1\end{pmatrix}=\begin{pmatrix}0 & 0\\ -1 & 1\end{pmatrix}: 2\times 2.$

$\mathbf{DE}=(1\quad 2)\begin{pmatrix}3\\ -1\end{pmatrix}=1 : 1\times 1.$

$\mathbf{ED}=\begin{pmatrix}3\\ -1\end{pmatrix}(1\quad 2)=\begin{pmatrix}3 & 6\\ -1 & -2\end{pmatrix}: 2\times 2.$

$\mathbf{DA}=(1\quad 2)\begin{pmatrix}1 & 2\\ 2 & 1\end{pmatrix}=(5\quad 4) : 1\times 2.$

AD not defined.
EA not defined.

$\mathbf{AE}=\begin{pmatrix}1 & 2\\ 2 & 1\end{pmatrix}\begin{pmatrix}3\\ -1\end{pmatrix}=\begin{pmatrix}1\\ 5\end{pmatrix}: 2\times 1.$

$\Rightarrow \mathbf{AB}-\mathbf{BA}=\begin{pmatrix}-1 & 3\\ 1 & 0\end{pmatrix}-\begin{pmatrix}-1 & 1\\ 3 & 0\end{pmatrix}=\begin{pmatrix}0 & 2\\ -2 & 0\end{pmatrix}: 2\times 2.$

$(\mathbf{AB})\mathbf{C}=\begin{pmatrix}-1 & 3\\ 1 & 0\end{pmatrix}\begin{pmatrix}-1 & 1\\ -1 & 1\end{pmatrix}=\begin{pmatrix}-2 & 2\\ -1 & 1\end{pmatrix}: 2\times 2.$

$\mathbf{A}(\mathbf{BC})=\begin{pmatrix}1 & 2\\ 2 & 1\end{pmatrix}\begin{pmatrix}0 & 0\\ -1 & 1\end{pmatrix}=\begin{pmatrix}-2 & 2\\ -1 & 1\end{pmatrix}: 2\times 2.$

$\mathbf{A}(\mathbf{B}+\mathbf{C})=\begin{pmatrix}1 & 2\\ 2 & 1\end{pmatrix}\begin{pmatrix}0 & 0\\ -2 & 3\end{pmatrix}=\begin{pmatrix}-4 & 6\\ -2 & 3\end{pmatrix}: 2\times 2.$

$\mathbf{AB}+\mathbf{AC}=\begin{pmatrix}-1 & 3\\ 1 & 0\end{pmatrix}+\begin{pmatrix}-3 & 3\\ -3 & 3\end{pmatrix}=\begin{pmatrix}-4 & 6\\ -2 & 3\end{pmatrix}: 2\times 2.$

4.5 (a) Reflection in the line $y = x$ will result in the x and y values interchanging. We can represent this coordinate transformation as:

$$\begin{pmatrix} x' \\ y' \end{pmatrix} = \begin{pmatrix} y \\ x \end{pmatrix} = \begin{pmatrix} d_{11} & d_{12} \\ d_{21} & d_{22} \end{pmatrix} \begin{pmatrix} x \\ y \end{pmatrix}$$
$$\mathbf{r}' \quad = \quad \mathbf{D} \quad \mathbf{r}$$

Multiplying out gives:

$$d_{11}x + d_{12}y = y \Rightarrow d_{11} = 0, \quad d_{12} = 1$$
$$d_{21}x + d_{22}y = x \Rightarrow d_{21} = 1, \quad d_{22} = 0$$
$$\Rightarrow \mathbf{D} = \begin{pmatrix} 0 & 1 \\ 1 & 0 \end{pmatrix}.$$

(b) (i) $\mathbf{E} = \mathbf{CD} = \begin{pmatrix} -1 & 0 \\ 0 & 1 \end{pmatrix} \begin{pmatrix} 0 & 1 \\ 1 & 0 \end{pmatrix} = \begin{pmatrix} 0 & -1 \\ 1 & 0 \end{pmatrix}.$

(ii) $\mathbf{F} = \mathbf{DC} = \begin{pmatrix} 0 & 1 \\ 1 & 0 \end{pmatrix} \begin{pmatrix} -1 & 0 \\ 0 & 1 \end{pmatrix} = \begin{pmatrix} 0 & 1 \\ -1 & 0 \end{pmatrix}.$

4.6 (a) $\mathbf{A}^{\mathrm{T}} = \begin{pmatrix} \cos\theta & \sin\theta \\ -\sin\theta & \cos\theta \end{pmatrix}.$

(b) $\mathbf{C}^{\mathrm{T}} = \begin{pmatrix} -1 & -1 \\ 1 & 1 \end{pmatrix}.$

(c) $\mathbf{D}^{\mathrm{T}} = \begin{pmatrix} 1 & 1 \\ 3 & 2 \\ 4 & 1 \end{pmatrix}.$

4.7 (a) $\mathbf{X}_{nm}\mathbf{X}_{mn}{}^{\mathrm{T}} = n \times n$; $\mathbf{X}_{mn}{}^{\mathrm{T}}\mathbf{X}_{nm} = m \times m$.

(b) $\mathbf{BB}^{\mathrm{T}} = \begin{pmatrix} 1 & 1 & 2 \\ 1 & 2 & 1 \end{pmatrix} \begin{pmatrix} 1 & 1 \\ 1 & 2 \\ 2 & 1 \end{pmatrix} = \begin{pmatrix} 6 & 5 \\ 5 & 6 \end{pmatrix};$

$\mathbf{B}^{\mathrm{T}}\mathbf{B} = \begin{pmatrix} 1 & 1 \\ 1 & 2 \\ 2 & 1 \end{pmatrix} \begin{pmatrix} 1 & 1 & 2 \\ 1 & 2 & 1 \end{pmatrix} = \begin{pmatrix} 2 & 3 & 3 \\ 3 & 5 & 4 \\ 3 & 4 & 5 \end{pmatrix}.$

4.8 $\mathbf{A} = \begin{pmatrix} 1+\mathrm{i} & \mathrm{i} \\ -\mathrm{i} & 1 \end{pmatrix}$; $\mathbf{A}^* = \begin{pmatrix} 1-\mathrm{i} & -\mathrm{i} \\ \mathrm{i} & 1 \end{pmatrix}$; $\mathbf{A}^{\dagger} = \begin{pmatrix} 1-\mathrm{i} & \mathrm{i} \\ -\mathrm{i} & 1 \end{pmatrix}.$

4.9 (a) $\mathbf{AB} = \begin{pmatrix} 1 & 1-\mathrm{i} \\ 1+\mathrm{i} & -1 \end{pmatrix} \begin{pmatrix} 1 & 1+\mathrm{i} \\ 1+\mathrm{i} & 0 \end{pmatrix} = \begin{pmatrix} 3 & 1+\mathrm{i} \\ 0 & 2\mathrm{i} \end{pmatrix};$

$(\mathbf{AB})^* = \begin{pmatrix} 3 & 1-\mathrm{i} \\ 0 & -2\mathrm{i} \end{pmatrix};$

$$\mathbf{A}^*\mathbf{B}^* = \begin{pmatrix} 1 & 1+i \\ 1-i & -1 \end{pmatrix}\begin{pmatrix} 1 & 1-i \\ 1-i & 0 \end{pmatrix} = \begin{pmatrix} 3 & 1-i \\ 0 & -2i \end{pmatrix}.$$

(b) $(\mathbf{AB})^\dagger = \begin{pmatrix} 3 & 0 \\ 1-i & -2i \end{pmatrix}$;

$$\mathbf{B}^\dagger\mathbf{A}^\dagger = \begin{pmatrix} 1 & 1-i \\ 1-i & 0 \end{pmatrix}\begin{pmatrix} 1 & 1-i \\ 1+i & -1 \end{pmatrix} = \begin{pmatrix} 3 & 0 \\ 1-i & -2i \end{pmatrix}.$$

4.10 (a) $\mathbf{AB} = \begin{pmatrix} 1 & -1 \\ 0 & 3 \end{pmatrix}\begin{pmatrix} 0 & 1 \\ 1 & -2 \end{pmatrix} = \begin{pmatrix} -1 & 3 \\ 3 & -6 \end{pmatrix}$

$\Rightarrow \mathrm{tr}(\mathbf{AB}) = -1 - 6 = -7$

$$\mathbf{BA} = \begin{pmatrix} 0 & 1 \\ 1 & -2 \end{pmatrix}\begin{pmatrix} 1 & -1 \\ 0 & 3 \end{pmatrix} = \begin{pmatrix} 0 & 3 \\ 1 & -7 \end{pmatrix} \Rightarrow \mathrm{tr}(\mathbf{BA}) = 0 - 7 = -7.$$

(b) $\mathbf{ABC} = \begin{pmatrix} -1 & 3 \\ 3 & -6 \end{pmatrix}\begin{pmatrix} -1 & 1 \\ 1 & 0 \end{pmatrix} = \begin{pmatrix} 4 & -1 \\ -9 & 3 \end{pmatrix} \Rightarrow \mathrm{tr}(\mathbf{ABC}) = 4 + 3 = 7.$

$$\mathbf{CAB} = \begin{pmatrix} -1 & 1 \\ 1 & 0 \end{pmatrix}\begin{pmatrix} -1 & 3 \\ 3 & -6 \end{pmatrix} - \begin{pmatrix} 4 & -9 \\ -1 & 3 \end{pmatrix}$$

$\Rightarrow \mathrm{tr}(\mathbf{CAB}) = 4 + 3 = 7$

$$\mathbf{BCA} = \begin{pmatrix} 0 & 1 \\ 1 & -2 \end{pmatrix}\begin{pmatrix} -1 & 4 \\ 1 & -1 \end{pmatrix} = \begin{pmatrix} 1 & -1 \\ -3 & 6 \end{pmatrix}$$

$\Rightarrow \mathrm{tr}(\mathbf{BCA}) = 1 + 6 = 7.$

(c) $\mathbf{D}^\mathrm{T}\mathbf{D} = \begin{pmatrix} 1 & 1 \\ -1 & -2 \\ 0 & 0 \end{pmatrix}\begin{pmatrix} 1 & -1 & 0 \\ 1 & -2 & 0 \end{pmatrix} - \begin{pmatrix} 2 & -3 & 0 \\ -3 & 5 & 0 \\ 0 & 0 & 0 \end{pmatrix}$

$\Rightarrow \mathrm{tr}(\mathbf{D}^\mathrm{T}\mathbf{D}) = 2 + 5 + 0 = 7.$

$$\mathbf{DD}^\mathrm{T} = \begin{pmatrix} 1 & -1 & 0 \\ 1 & -2 & 0 \end{pmatrix}\begin{pmatrix} 1 & 1 \\ -1 & -2 \\ 0 & 0 \end{pmatrix} = \begin{pmatrix} 2 & 3 \\ 3 & 5 \end{pmatrix}$$

$\Rightarrow \mathrm{tr}(\mathbf{DD}^\mathrm{T}) = 2 + 5 = 7.$

4.11 (a) $\begin{pmatrix} 0 & 0 & 0 \\ 0 & 0 & 0 \end{pmatrix}\begin{pmatrix} 1 & 3 \\ 2 & 2 \\ 0 & 1 \end{pmatrix} = \begin{pmatrix} 0 & 0 \\ 0 & 0 \end{pmatrix}.$

(b) $\begin{pmatrix} 1 & 2 & 3 \\ 4 & 5 & 6 \end{pmatrix}\begin{pmatrix} 0 & 0 & 0 \\ 0 & 0 & 0 \end{pmatrix}$ is undefined.

(c) $\begin{pmatrix} 1 & 0 & 0 \\ 0 & 1 & 0 \\ 0 & 0 & 1 \end{pmatrix}\begin{pmatrix} 1 & 3 \\ 2 & 2 \\ 0 & 1 \end{pmatrix} = \begin{pmatrix} 1 & 3 \\ 2 & 2 \\ 0 & 1 \end{pmatrix}$.

(d) $\begin{pmatrix} 1 & 2 & 3 \\ 4 & 5 & 6 \end{pmatrix}\begin{pmatrix} 1 & 0 & 0 \\ 0 & 1 & 0 \\ 0 & 0 & 1 \end{pmatrix} = \begin{pmatrix} 1 & 2 & 3 \\ 4 & 5 & 6 \end{pmatrix}$.

(e) $\begin{pmatrix} 1 & 3 \\ 2 & 2 \\ 0 & 1 \end{pmatrix}\begin{pmatrix} 1 & 0 & 0 \\ 0 & 1 & 0 \\ 0 & 0 & 1 \end{pmatrix}$ is undefined.

4.12 (a) (i) $\det \mathbf{A} = \begin{vmatrix} \cos\theta & -\sin\theta \\ \sin\theta & \cos\theta \end{vmatrix} = \cos^2\theta + \sin^2\theta = 1;$

$$\mathbf{B} = \begin{pmatrix} \cos\theta & -\sin\theta \\ \sin\theta & \cos\theta \end{pmatrix}$$

(ii) $\mathbf{B}^{\mathrm{T}}\mathbf{A} = \begin{pmatrix} \cos\theta & \sin\theta \\ -\sin\theta & \cos\theta \end{pmatrix}\begin{pmatrix} \cos\theta & -\sin\theta \\ \sin\theta & \cos\theta \end{pmatrix} = \begin{pmatrix} 1 & 0 \\ 0 & 1 \end{pmatrix}$.

$\mathbf{E}_2 \det \mathbf{A} = \begin{pmatrix} 1 & 0 \\ 0 & 1 \end{pmatrix} \times 1 = \begin{pmatrix} 1 & 0 \\ 0 & 1 \end{pmatrix}$. Thus $\mathbf{E}_2 \det \mathbf{A} = \mathbf{B}^{\mathrm{T}}\mathbf{A}$.

(b) $G_{11} = \begin{pmatrix} -1 & 2 \\ 1 & 1 \end{pmatrix} = -3;\ G_{12} = -\begin{pmatrix} 2 & 2 \\ 1 & 1 \end{pmatrix} = 0;\ G_{13} = \begin{pmatrix} 2 & -1 \\ 1 & 1 \end{pmatrix} = 3;$

$G_{21} = -\begin{pmatrix} -1 & 1 \\ 1 & 1 \end{pmatrix} = 2;\ G_{22} = \begin{pmatrix} 0 & 1 \\ 1 & 1 \end{pmatrix} = -1;\ G_{23} = -\begin{pmatrix} 0 & -1 \\ 1 & 1 \end{pmatrix} = -1;$

$G_{31} = \begin{pmatrix} -1 & 1 \\ -1 & 2 \end{pmatrix} = -1;\ G_{32} = -\begin{pmatrix} 0 & 1 \\ 2 & 2 \end{pmatrix} = 2;\ G_{33} = \begin{pmatrix} 0 & -1 \\ 2 & -1 \end{pmatrix} = 2;$

$$\mathbf{H} = \begin{pmatrix} -3 & 0 & 3 \\ 2 & -1 & -1 \\ -1 & 2 & 2 \end{pmatrix} \Rightarrow \mathbf{H}^{\mathrm{T}}\mathbf{G} = \begin{pmatrix} -3 & 2 & -1 \\ 0 & -1 & 2 \\ 3 & -1 & 2 \end{pmatrix}\begin{pmatrix} 0 & -1 & 1 \\ 2 & -1 & 2 \\ 1 & 1 & 1 \end{pmatrix}$$

$$= \begin{pmatrix} 3 & 0 & 0 \\ 0 & 3 & 0 \\ 0 & 0 & 3 \end{pmatrix}.$$

$\det \mathbf{G} = 1\begin{vmatrix} 2 & 2 \\ 1 & 1 \end{vmatrix} + 1\begin{vmatrix} 2 & -1 \\ 1 & 1 \end{vmatrix} = 0 + 3 = 3;$ therefore

$$\mathbf{E}_3 \det \mathbf{G} = \begin{pmatrix} 1 & 0 & 0 \\ 0 & 1 & 0 \\ 0 & 0 & 1 \end{pmatrix} \times 3 = \begin{pmatrix} 3 & 0 & 0 \\ 0 & 3 & 0 \\ 0 & 0 & 3 \end{pmatrix}.$$

4.13 (a) $\mathbf{A} = \begin{pmatrix} \frac{1}{\sqrt{2}} & k \\ \frac{1}{\sqrt{2}} & -\frac{1}{\sqrt{2}} \end{pmatrix}$; $\det \mathbf{A} = -\dfrac{1}{2} - \dfrac{k}{\sqrt{2}} = \pm 1$

$$\Rightarrow -\frac{k}{\sqrt{2}} = \pm 1 + \frac{1}{2} \Rightarrow \frac{k}{\sqrt{2}} = \mp\, 1 - \frac{1}{2} = -\frac{3}{2}, \frac{1}{2}$$

$$\Rightarrow k = -\frac{3\sqrt{2}}{2}, \frac{\sqrt{2}}{2} = -\frac{3}{\sqrt{2}}, \frac{1}{\sqrt{2}}.$$

Check: $\mathbf{A}^{\mathrm{T}}\mathbf{A} = \mathbf{A}\mathbf{A}^{\mathrm{T}} = \mathbf{E}_n \Rightarrow \mathbf{A}^{\mathrm{T}}\mathbf{A} = \begin{pmatrix} \frac{1}{\sqrt{2}} & \frac{1}{\sqrt{2}} \\ k & -\frac{1}{\sqrt{2}} \end{pmatrix}\begin{pmatrix} \frac{1}{\sqrt{2}} & k \\ \frac{1}{\sqrt{2}} & -\frac{1}{\sqrt{2}} \end{pmatrix}$

$$= \begin{pmatrix} 1 & \frac{k}{\sqrt{2}} - \frac{1}{2} \\ \frac{k}{\sqrt{2}} - \frac{1}{2} & k^2 + \frac{1}{2} \end{pmatrix}$$

$$\mathbf{A}\mathbf{A}^{\mathrm{T}} = \begin{pmatrix} \frac{1}{\sqrt{2}} & k \\ \frac{1}{\sqrt{2}} & -\frac{1}{\sqrt{2}} \end{pmatrix}\begin{pmatrix} \frac{1}{\sqrt{2}} & \frac{1}{\sqrt{2}} \\ k & -\frac{1}{\sqrt{2}} \end{pmatrix} = \begin{pmatrix} \frac{1}{2} + k^2 & \frac{1}{2} - \frac{k}{\sqrt{2}} \\ \frac{1}{2} - \frac{k}{\sqrt{2}} & 1 \end{pmatrix}$$

For $k = -\dfrac{3}{\sqrt{2}}, \mathbf{A}^{\mathrm{T}}\mathbf{A} = \begin{pmatrix} 1 & -2 \\ -2 & 5 \end{pmatrix}; \mathbf{A}\mathbf{A}^{\mathrm{T}} = \begin{pmatrix} 5 & 2 \\ 2 & 1 \end{pmatrix} \Rightarrow \mathbf{A}^{\mathrm{T}}\mathbf{A} \neq \mathbf{A}\mathbf{A}^{\mathrm{T}} \neq \mathbf{E}_n.$

For $k = \dfrac{1}{\sqrt{2}}, \mathbf{A}^{\mathrm{T}}\mathbf{A} = \begin{pmatrix} 1 & 0 \\ 0 & 1 \end{pmatrix}; \mathbf{A}\mathbf{A}^{\mathrm{T}} = \begin{pmatrix} 1 & 0 \\ 0 & 1 \end{pmatrix} \Rightarrow \mathbf{A}^{\mathrm{T}}\mathbf{A} = \mathbf{A}\mathbf{A}^{\mathrm{T}} = \mathbf{E}_n.$

The only valid solution is $k = \frac{1}{\sqrt{2}}$.

(b) $\mathbf{R} = \begin{pmatrix} \cos\theta & \sin\theta & 0 \\ \sin\theta & \cos\theta & 0 \\ 0 & 0 & 1 \end{pmatrix}$; $\det \mathbf{R} = \cos^2\theta - \sin^2\theta = +1.$

However, $\cos 2\theta = \cos^2\theta - \sin^2\theta$ and so $\cos 2\theta = \pm 1$

$$\Rightarrow 2\theta = \cos^{-1} \pm 1 = 0, \pm n\pi \Rightarrow \theta = 0, \pm\frac{n\pi}{2}, n = 1, 2, 3, \ldots$$

$$\Rightarrow \mathbf{R} = \begin{pmatrix} 1 & 0 & 0 \\ 0 & 1 & 0 \\ 0 & 0 & 1 \end{pmatrix}, \text{ for } \theta = 0; \mathbf{R} = \begin{pmatrix} 0 & 1 & 0 \\ 1 & 0 & 0 \\ 0 & 0 & 1 \end{pmatrix}, \text{ for } \theta = \frac{\pi}{2};$$

$$\mathbf{R} = \begin{pmatrix} -1 & 0 & 0 \\ 0 & -1 & 0 \\ 0 & 0 & 1 \end{pmatrix}, \text{ for } \theta = \pi; \mathbf{R} = \begin{pmatrix} 0 & -1 & 0 \\ -1 & 0 & 0 \\ 0 & 0 & 1 \end{pmatrix}, \text{ for } \theta = \frac{3\pi}{2}, -\frac{\pi}{2}.$$

In each case, $\mathbf{R}^T = \mathbf{R}$.

$$\mathbf{RR}^T = \mathbf{R}^T\mathbf{R} = \begin{pmatrix} 1 & 0 & 0 \\ 0 & 1 & 0 \\ 0 & 0 & 1 \end{pmatrix}\begin{pmatrix} 1 & 0 & 0 \\ 0 & 1 & 0 \\ 0 & 0 & 1 \end{pmatrix} = \begin{pmatrix} 1 & 0 & 0 \\ 0 & 1 & 0 \\ 0 & 0 & 1 \end{pmatrix} = \mathbf{E}_3$$

$$\mathbf{RR}^T = \mathbf{R}^T\mathbf{R} = \begin{pmatrix} 0 & 1 & 0 \\ 1 & 0 & 0 \\ 0 & 0 & 1 \end{pmatrix}\begin{pmatrix} 0 & 1 & 0 \\ 1 & 0 & 0 \\ 0 & 0 & 1 \end{pmatrix} = \begin{pmatrix} 1 & 0 & 0 \\ 0 & 1 & 0 \\ 0 & 0 & 1 \end{pmatrix} = \mathbf{E}_3$$

$$\mathbf{RR}^T = \mathbf{R}^T\mathbf{R} = \begin{pmatrix} -1 & 0 & 0 \\ 0 & -1 & 0 \\ 0 & 0 & 1 \end{pmatrix}\begin{pmatrix} -1 & 0 & 0 \\ 0 & -1 & 0 \\ 0 & 0 & 1 \end{pmatrix} = \begin{pmatrix} 1 & 0 & 0 \\ 0 & 1 & 0 \\ 0 & 0 & 1 \end{pmatrix} = \mathbf{E}_3$$

$$\mathbf{RR}^T = \mathbf{R}^T\mathbf{R} = \begin{pmatrix} 0 & -1 & 0 \\ -1 & 0 & 0 \\ 0 & 0 & 1 \end{pmatrix}\begin{pmatrix} 0 & -1 & 0 \\ -1 & 0 & 0 \\ 0 & 0 & 1 \end{pmatrix} = \begin{pmatrix} 1 & 0 & 0 \\ 0 & 1 & 0 \\ 0 & 0 & 1 \end{pmatrix} = \mathbf{E}_3$$

For $\theta = 0, \ \pm\frac{n\pi}{2}, n = 1, 2, 3, \ldots, \mathbf{R}^T\mathbf{R} = \mathbf{RR}^T = \mathbf{E_n}$.

(c) $\det \mathbf{A} = 1 + k \Rightarrow 1 + k = \pm 1 \Rightarrow k = 0, -2$

For $k = 0$:

$$\mathbf{AA}^T = \begin{pmatrix} 1 & 0 \\ -1 & 1 \end{pmatrix}\begin{pmatrix} 1 & -1 \\ 0 & 1 \end{pmatrix} = \begin{pmatrix} 1 & -1 \\ -1 & 2 \end{pmatrix}$$

$$\mathbf{A}^T\mathbf{A} = \begin{pmatrix} 1 & -1 \\ 0 & 1 \end{pmatrix}\begin{pmatrix} 1 & 0 \\ -1 & 1 \end{pmatrix} = \begin{pmatrix} 2 & -1 \\ -1 & 1 \end{pmatrix}$$

For $k = -2$:

$$\mathbf{AA}^T = \begin{pmatrix} 1 & -2 \\ -1 & 1 \end{pmatrix}\begin{pmatrix} 1 & -1 \\ -2 & 1 \end{pmatrix} = \begin{pmatrix} 5 & -3 \\ -3 & 2 \end{pmatrix}$$

$$\mathbf{A}^T\mathbf{A} = \begin{pmatrix} 1 & -1 \\ -2 & 1 \end{pmatrix}\begin{pmatrix} 1 & -2 \\ -1 & 1 \end{pmatrix} = \begin{pmatrix} 2 & -3 \\ -3 & 5 \end{pmatrix}$$

$\therefore$ Not orthogonal.

4.14. (a) $\mathbf{A} = \begin{pmatrix} 0 & 3+\mathrm{i} \\ 3-\mathrm{i} & 1 \end{pmatrix}$, $\mathbf{A}^\dagger = \begin{pmatrix} 0 & 3+\mathrm{i} \\ 3-\mathrm{i} & 1 \end{pmatrix}$, $\therefore \mathbf{A} = \mathbf{A}^\dagger$.

(b) $\mathbf{Ax} = \begin{pmatrix} 0 & 3+\mathrm{i} \\ 3-\mathrm{i} & 1 \end{pmatrix}\begin{pmatrix} 1 \\ \mathrm{i} \end{pmatrix} = \begin{pmatrix} 3\mathrm{i}-1 \\ 3 \end{pmatrix}$; $\mathbf{x}^\dagger\mathbf{Ax}$

$$= (1 \quad -\mathrm{i})\begin{pmatrix} 3\mathrm{i}-1 \\ 3 \end{pmatrix} = 3\mathrm{i} - 1 - 3\mathrm{i} = -1.$$

4.15. (a) $\mathbf{A} = \begin{pmatrix} 1 & \mathrm{i} \\ -\mathrm{i} & 0 \end{pmatrix}$; $\mathbf{A}^\dagger = \begin{pmatrix} 1 & \mathrm{i} \\ -\mathrm{i} & 0 \end{pmatrix}$ $\therefore$ Hermitian.

$$\mathbf{AA}^\dagger = \begin{pmatrix} 1 & \mathrm{i} \\ -\mathrm{i} & 0 \end{pmatrix}\begin{pmatrix} 1 & \mathrm{i} \\ -\mathrm{i} & 0 \end{pmatrix} = \begin{pmatrix} 2 & \mathrm{i} \\ -\mathrm{i} & 1 \end{pmatrix} \therefore \text{ Not unitary.}$$

(b) $\mathbf{B} = \dfrac{1}{\sqrt{2}}\begin{pmatrix} 1 & \mathrm{i} \\ -\mathrm{i} & -1 \end{pmatrix}$; $\mathbf{B}^\dagger = \dfrac{1}{\sqrt{2}}\begin{pmatrix} 1 & \mathrm{i} \\ -\mathrm{i} & -1 \end{pmatrix}$ $\therefore$ Hermitian.

$$\mathbf{BB}^\dagger = \frac{1}{2}\begin{pmatrix} 1 & \mathrm{i} \\ -\mathrm{i} & -1 \end{pmatrix}\begin{pmatrix} 1 & \mathrm{i} \\ -\mathrm{i} & -1 \end{pmatrix} = \frac{1}{2}\begin{pmatrix} 2 & 0 \\ 0 & 2 \end{pmatrix} = \begin{pmatrix} 1 & 0 \\ 0 & 1 \end{pmatrix} \therefore \text{ Unitary.}$$

(c) $\mathbf{C} = \begin{pmatrix} 0 & -1 \\ 1 & 0 \end{pmatrix}$; $\mathbf{C}^\mathrm{T} = \begin{pmatrix} 0 & 1 \\ -1 & 0 \end{pmatrix}$ $\therefore$ Not symmetric.

$$\mathbf{CC}^\mathrm{T} = \begin{pmatrix} 0 & -1 \\ 1 & 0 \end{pmatrix}\begin{pmatrix} 0 & 1 \\ -1 & 0 \end{pmatrix} = \begin{pmatrix} 1 & 0 \\ 0 & 1 \end{pmatrix} \therefore \text{ Orthogonal.}$$

(d) $\mathbf{D} = \begin{pmatrix} 1 & -1 \\ -1 & 0 \end{pmatrix}$; $\mathbf{D}^\mathrm{T} = \begin{pmatrix} 1 & -1 \\ -1 & 0 \end{pmatrix}$ $\therefore$ Symmetric.

$$\mathbf{DD}^\mathrm{T} = \begin{pmatrix} 1 & -1 \\ -1 & 0 \end{pmatrix}\begin{pmatrix} 1 & -1 \\ -1 & 0 \end{pmatrix} = \begin{pmatrix} 2 & -1 \\ -1 & 1 \end{pmatrix}$$

$\therefore$ Not orthogonal.

4.16 $\det \mathbf{A} = \begin{vmatrix} 1 & -1 & 1 \\ -1 & -1 & 1 \\ 1 & 1 & 1 \end{vmatrix} \Rightarrow$ add col 2 to col 3 $\Rightarrow \begin{vmatrix} 1 & -1 & 0 \\ -1 & -1 & 0 \\ 1 & 1 & 2 \end{vmatrix}$

and then add col 1 to col 2 $\Rightarrow \det \mathbf{A} = \begin{vmatrix} 1 & 0 & 0 \\ -1 & -2 & 0 \\ 1 & 2 & 2 \end{vmatrix} = \begin{vmatrix} -2 & 0 \\ 2 & 2 \end{vmatrix} = -4.$

The matrix of cofactors of $\mathbf{A}$ is $\mathbf{B} = \begin{pmatrix} -2 & 2 & 0 \\ 2 & 0 & -2 \\ 0 & -2 & -2 \end{pmatrix}$;

$$\mathbf{B}^\mathrm{T} = \begin{pmatrix} -2 & 2 & 0 \\ 2 & 0 & -2 \\ 0 & -2 & -2 \end{pmatrix}$$

Therefore $\mathbf{A}^{-1} = \dfrac{1}{\det \mathbf{A}}\mathbf{B}^\mathrm{T} = -\dfrac{1}{4}\begin{pmatrix} -2 & 2 & 0 \\ 2 & 0 & -2 \\ 0 & -2 & -2 \end{pmatrix}$

$$= \begin{pmatrix} \frac{1}{2} & -\frac{1}{2} & 0 \\ -\frac{1}{2} & 0 & \frac{1}{2} \\ 0 & \frac{1}{2} & \frac{1}{2} \end{pmatrix}.$$

Check : $\mathbf{AA}^{-1} = \begin{pmatrix} 1 & -1 & 0 \\ -1 & -1 & 1 \\ 1 & 1 & 1 \end{pmatrix}\begin{pmatrix} \frac{1}{2} & -\frac{1}{2} & 0 \\ -\frac{1}{2} & 0 & \frac{1}{2} \\ 0 & \frac{1}{2} & \frac{1}{2} \end{pmatrix} = \begin{pmatrix} 1 & 0 & 0 \\ 0 & 1 & 0 \\ 0 & 0 & 1 \end{pmatrix}$.

4.17. $x + 2y + 3z = 1$

$$8y + 2z = 1$$

$$-2x + 4y + 2z = 2$$

$$\underset{\mathbf{A}}{\begin{pmatrix} 1 & 2 & 3 \\ 0 & 8 & 2 \\ -2 & 4 & 2 \end{pmatrix}}\underset{\mathbf{x}}{\begin{pmatrix} x \\ y \\ z \end{pmatrix}} = \underset{\mathbf{b}}{\begin{pmatrix} 1 \\ 1 \\ 2 \end{pmatrix}}$$

Matrix of cofactors $\mathbf{B} = \begin{pmatrix} 8 & -4 & 16 \\ 8 & 8 & -8 \\ -20 & -2 & 8 \end{pmatrix}$; $\mathbf{B}^{\mathrm{T}} = \begin{pmatrix} 8 & 8 & -20 \\ -4 & 8 & -2 \\ 16 & -8 & 8 \end{pmatrix}$;

$$\det \mathbf{A} = \begin{vmatrix} 8 & 2 \\ 4 & 2 \end{vmatrix} - 2\begin{vmatrix} 0 & 2 \\ -2 & 2 \end{vmatrix} + 3\begin{vmatrix} 0 & 8 \\ -2 & 4 \end{vmatrix} = 8 - 8 + 48 = 48;$$

$$\mathbf{A}^{-1} = \frac{1}{\det \mathbf{A}}\mathbf{B}^{\mathrm{T}} = \frac{1}{48}\begin{pmatrix} 8 & 8 & -20 \\ -4 & 8 & -2 \\ 16 & -8 & 8 \end{pmatrix} = \begin{pmatrix} \frac{1}{6} & \frac{1}{6} & -\frac{5}{12} \\ -\frac{1}{12} & \frac{1}{6} & -\frac{1}{24} \\ \frac{1}{3} & -\frac{1}{6} & \frac{1}{6} \end{pmatrix};$$

$$\begin{pmatrix} x \\ y \\ z \end{pmatrix} = \begin{pmatrix} \frac{1}{6} & \frac{1}{6} & -\frac{5}{12} \\ -\frac{1}{12} & \frac{1}{6} & -\frac{1}{24} \\ \frac{1}{3} & -\frac{1}{6} & \frac{1}{6} \end{pmatrix}\begin{pmatrix} 1 \\ 1 \\ 2 \end{pmatrix} = \begin{pmatrix} -\frac{1}{2} \\ 0 \\ \frac{1}{2} \end{pmatrix} \Rightarrow x = -\frac{1}{2},\ y = 0,\ z = \frac{1}{2}.$$

4.18. $\begin{pmatrix} (\alpha - \varepsilon) & \beta & 0 \\ \beta & (\alpha - \varepsilon) & \beta \\ 0 & \beta & (\alpha - \varepsilon) \end{pmatrix}\begin{pmatrix} c_1 \\ c_2 \\ c_3 \end{pmatrix} = \mathbf{0}$:

(a) For $\varepsilon = \alpha$, $\begin{pmatrix} 0 & \beta & 0 \\ \beta & 0 & \beta \\ 0 & \beta & 0 \end{pmatrix}\begin{pmatrix} c_1 \\ c_2 \\ c_3 \end{pmatrix} = \begin{pmatrix} c_2\beta \\ c_1\beta + c_3\beta \\ c_2\beta \end{pmatrix} = \begin{pmatrix} 0 \\ 0 \\ 0 \end{pmatrix}$

$$\Rightarrow c_2\beta = 0;\ c_1\beta + c_3\beta = 0 \quad \Rightarrow \quad c_2 = 0,\ c_3 = -c_1$$

(b) For $\varepsilon = \alpha + \sqrt{2}\beta$, $\begin{pmatrix} -\sqrt{2}\beta & \beta & 0 \\ \beta & -\sqrt{2}\beta & \beta \\ 0 & \beta & -\sqrt{2}\beta \end{pmatrix}\begin{pmatrix} c_1 \\ c_2 \\ c_3 \end{pmatrix}$

$$= \begin{pmatrix} -\sqrt{2}\beta c_1 + c_2\beta \\ c_1\beta - \sqrt{2}\beta c_2 + c_3\beta \\ c_2\beta - \sqrt{2}\beta c_3 \end{pmatrix} = \begin{pmatrix} 0 \\ 0 \\ 0 \end{pmatrix}$$

$\Rightarrow -\sqrt{2}\beta c_1 + c_2\beta = 0;\ c_1\beta - \sqrt{2}\beta c_2 + c_3\beta = 0;\ c_2\beta - \sqrt{2}\beta c_3 = 0$
$\Rightarrow c_2 = \sqrt{2}c_1,\ c_3 = c_1.$

(c) For $\varepsilon = \alpha - \sqrt{2}\beta$, $\begin{pmatrix} \sqrt{2}\beta & \beta & 0 \\ \beta & \sqrt{2}\beta & \beta \\ 0 & \beta & \sqrt{2}\beta \end{pmatrix}\begin{pmatrix} c_1 \\ c_2 \\ c_3 \end{pmatrix}$

$$= \begin{pmatrix} \sqrt{2}\beta c_1 + c_2\beta \\ c_1\beta + \sqrt{2}\beta c_2 + c_3\beta \\ c_2\beta + \sqrt{2}\beta c_3 \end{pmatrix} = \begin{pmatrix} 0 \\ 0 \\ 0 \end{pmatrix}$$

$\Rightarrow \sqrt{2}\beta c_1 + c_2\beta = 0;\ c_1\beta + \sqrt{2}\beta c_2 + c_3\beta = 0;\ c_2\beta + \sqrt{2}\beta c_3 = 0$
$\Rightarrow c_2 = -\sqrt{2}c_1,\ c_3 = c_1.$

(d) For $\varepsilon = \alpha$, $\mathbf{c} = c_1\begin{pmatrix} 1 \\ 0 \\ -1 \end{pmatrix}$; for $\varepsilon = \alpha + \sqrt{2}\beta$, $\mathbf{c} = c_1\begin{pmatrix} 1 \\ \sqrt{2} \\ 1 \end{pmatrix}$;

for $\varepsilon = \alpha - \sqrt{2}\beta$, $\mathbf{c} = c_1\begin{pmatrix} 1 \\ -\sqrt{2} \\ 1 \end{pmatrix}$

4.19. Identity is 1.

$$\left.\begin{array}{r} -1 \times -1 = 1 \\ -1 \times \mathrm{i} = -\mathrm{i} \\ -1 \times -\mathrm{i} = \mathrm{i} \\ \mathrm{i} \times \mathrm{i} = -1 \\ \mathrm{i} \times -\mathrm{i} = 1 \\ -\mathrm{i} \times -\mathrm{i} = -1 \end{array}\right\} \text{Product of any two yields another member of the group.}$$

Inverse of 1 is 1.
Inverse of -1 is -1.
Inverse of i is $-$i.
Inverse of $-$i is i.

Multiplication is associative: $(-1 \times \mathrm{i}) \times -\mathrm{i} = -1 \times (\mathrm{i} \times -\mathrm{i}) = -1$.

$\therefore$ The set G_1 forms a group.

4.20. (a)

	A	B	C	D
A	A	B	C	D
B	B	A	D	C
C	C	D	B	A
D	D	C	A	B

(b)

	A	B	C	D
A	A	B	C	D
B	B	A	D	C
C	C	D	A	B
D	D	C	B	A

4.21. One three-fold axis of rotation, three two-fold axes of rotation, one mirror plane containing the plane of the molecule, and three mirror planes perpendicular to the plane of the molecule.

Chapter 5

5.1. (a)

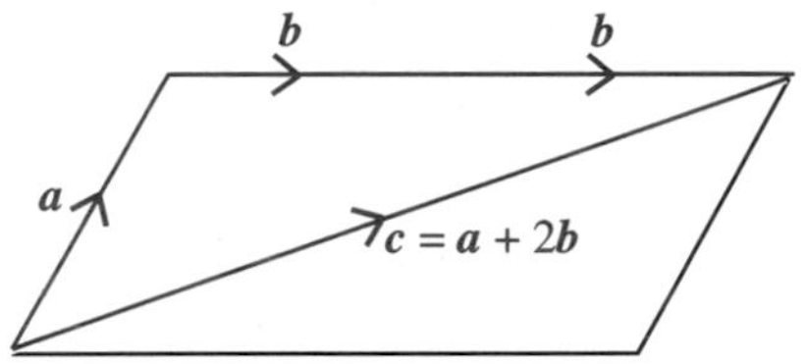

(b)

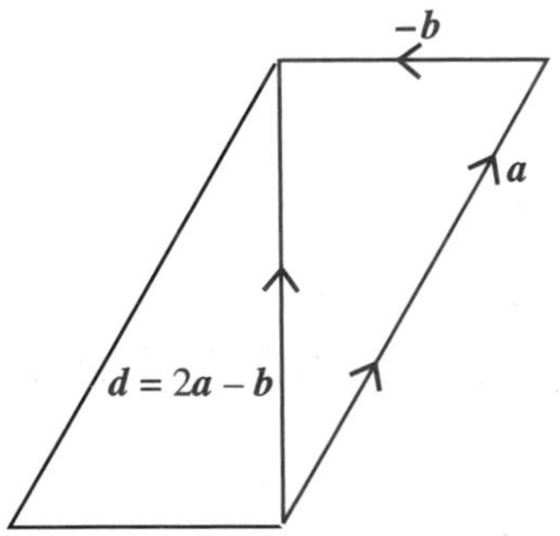

5.2. (a) $3a - 2b = 3(\hat{\boldsymbol{i}} + \hat{\boldsymbol{j}} - 2\hat{\boldsymbol{k}}) - 2(\hat{\boldsymbol{i}} + \hat{\boldsymbol{k}}) = \hat{\boldsymbol{i}} + 3\hat{\boldsymbol{j}} - 8\hat{\boldsymbol{k}}$.

(b) $-2\boldsymbol{a} - \boldsymbol{b} = -2(\hat{\boldsymbol{i}} + \hat{\boldsymbol{j}} - 2\hat{\boldsymbol{k}}) - (\hat{\boldsymbol{i}} + \hat{\boldsymbol{k}}) = -3\hat{\boldsymbol{i}} - 2\hat{\boldsymbol{j}} + 3\hat{\boldsymbol{k}}$.

(c) $\boldsymbol{a} + \boldsymbol{b} - \boldsymbol{c} - \boldsymbol{d} = (\hat{\boldsymbol{i}} + \hat{\boldsymbol{j}} - 2\hat{\boldsymbol{k}}) + (\hat{\boldsymbol{i}} + \hat{\boldsymbol{k}}) - (\hat{\boldsymbol{i}} + \hat{\boldsymbol{j}} + \hat{\boldsymbol{k}}) - (\hat{\boldsymbol{i}} - 2\hat{\boldsymbol{k}})$

$$= 0\hat{\boldsymbol{i}} + 0\hat{\boldsymbol{j}} + 0\hat{\boldsymbol{k}} = \boldsymbol{0}.$$

(d) $|\boldsymbol{a} - \boldsymbol{d}| = \left|(\hat{\boldsymbol{i}} + \hat{\boldsymbol{j}} - 2\hat{\boldsymbol{k}}) - (\hat{\boldsymbol{i}} - 2\hat{\boldsymbol{k}})\right| = \left|\hat{\boldsymbol{j}}\right| = 1.$

(e) $$\frac{\boldsymbol{a} + \boldsymbol{c}}{|\boldsymbol{a} + \boldsymbol{c}|} = \frac{(\hat{\boldsymbol{i}} + \hat{\boldsymbol{j}} - 2\hat{\boldsymbol{k}}) + (\hat{\boldsymbol{i}} + \hat{\boldsymbol{j}} + \hat{\boldsymbol{k}})}{\left|(\hat{\boldsymbol{i}} + \hat{\boldsymbol{j}} - 2\hat{\boldsymbol{k}}) + (\hat{\boldsymbol{i}} + \hat{\boldsymbol{j}} + \hat{\boldsymbol{k}})\right|} = \frac{2\hat{\boldsymbol{i}} + 2\hat{\boldsymbol{j}} - \hat{\boldsymbol{k}}}{\sqrt{2^2 + 2^2 + 1^2}}$$

$$= \frac{2\hat{\boldsymbol{i}} + 2\hat{\boldsymbol{j}} - \hat{\boldsymbol{k}}}{3} = \frac{2}{3}\hat{\boldsymbol{i}} + \frac{2}{3}\hat{\boldsymbol{j}} - \frac{1}{3}\hat{\boldsymbol{k}}.$$

(f) $$\left|\frac{2}{3}\hat{\boldsymbol{i}} + \frac{2}{3}\hat{\boldsymbol{j}} - \frac{1}{3}\hat{\boldsymbol{k}}\right| = \sqrt{\left(\tfrac{2}{3}\right)^2 + \left(\tfrac{2}{3}\right)^2 + \left(\tfrac{1}{3}\right)^2} = 1.$$

(g) $|\boldsymbol{a}| - |\boldsymbol{c}| = \left|(\hat{\boldsymbol{i}} + \hat{\boldsymbol{j}} - 2\hat{\boldsymbol{k}})\right| - \left|(\hat{\boldsymbol{i}} + \hat{\boldsymbol{j}} + \hat{\boldsymbol{k}})\right|$

$$= \sqrt{1 + 1 + 4} - \sqrt{1 + 1 + 1} = \sqrt{6} - \sqrt{3}.$$

5.3. (a) $\hat{\boldsymbol{i}},\ -\hat{\boldsymbol{j}},\ -\hat{\boldsymbol{i}},\ \hat{\boldsymbol{j}}$.

(b) $R\hat{\boldsymbol{i}},\ -R\hat{\boldsymbol{j}},\ -R\hat{\boldsymbol{i}},\ R\hat{\boldsymbol{j}}$.

(c) Shortest: e.g. $\boldsymbol{r} = R\hat{\boldsymbol{i}} - R\hat{\boldsymbol{j}};\ |\boldsymbol{r}| = \left|R\hat{\boldsymbol{i}} - R\hat{\boldsymbol{j}}\right| = \sqrt{R^2 + R^2} = \sqrt{2}R$.

Longest: e.g. $\boldsymbol{p} = -R\hat{\boldsymbol{i}} + -R\hat{\boldsymbol{i}} = -2R\hat{\boldsymbol{i}};\ |\boldsymbol{p}| = \left|-2R\hat{\boldsymbol{i}}\right| = 2R$.

5.4. (a) $\boldsymbol{a} \cdot \boldsymbol{c} = (2\hat{\boldsymbol{i}} + 3\hat{\boldsymbol{k}}) \cdot (\hat{\boldsymbol{i}} - 2\hat{\boldsymbol{j}} + \hat{\boldsymbol{k}}) = 2\hat{\boldsymbol{i}} \cdot \hat{\boldsymbol{i}} + 3\hat{\boldsymbol{k}} \cdot \hat{\boldsymbol{k}} = 5$.

(b) $\boldsymbol{a} \cdot (\boldsymbol{b} - 2\boldsymbol{c}) = (2\hat{\boldsymbol{i}} + 3\hat{\boldsymbol{k}}) \cdot (-\hat{\boldsymbol{i}} + 5\hat{\boldsymbol{j}} - \hat{\boldsymbol{k}}) = -2\hat{\boldsymbol{i}} \cdot \hat{\boldsymbol{i}} - 3\hat{\boldsymbol{k}} \cdot \hat{\boldsymbol{k}} = -5$.

(c) $\boldsymbol{a} \cdot (\boldsymbol{b} + \boldsymbol{a}) = (2\hat{\boldsymbol{i}} + 3\hat{\boldsymbol{k}}) \cdot (3\hat{\boldsymbol{i}} + \hat{\boldsymbol{j}} + 4\hat{\boldsymbol{k}}) = 6\hat{\boldsymbol{i}} \cdot \hat{\boldsymbol{i}} + 12\hat{\boldsymbol{k}} \cdot \hat{\boldsymbol{k}} = 18$.

(d) $\boldsymbol{b} \cdot \boldsymbol{c} = (\hat{\boldsymbol{i}} + \hat{\boldsymbol{j}} + \hat{\boldsymbol{k}}) \cdot (\hat{\boldsymbol{i}} - 2\hat{\boldsymbol{j}} + \hat{\boldsymbol{k}}) = \hat{\boldsymbol{i}} \cdot \hat{\boldsymbol{i}} - 2\hat{\boldsymbol{j}} \cdot \hat{\boldsymbol{j}} + \hat{\boldsymbol{k}} \cdot \hat{\boldsymbol{k}} = 0$.

5.5. (a) (i) $\boldsymbol{a} \cdot (\boldsymbol{b} - 2\boldsymbol{c}) = -5 = |\boldsymbol{a}| \cdot |(\boldsymbol{b} - 2\boldsymbol{c})| \cos\theta$

$$|\boldsymbol{a}| = \sqrt{2^2 + 3^2} = \sqrt{13}.$$

$$|(\boldsymbol{b} - 2\boldsymbol{c})| = \sqrt{1^2 + 5^2 + 1^2} = \sqrt{27}$$

$$\Rightarrow \frac{-5}{\sqrt{13}\sqrt{27}} = \cos\theta \Rightarrow \theta = 105.48^\circ.$$

(ii) $\boldsymbol{b} \cdot \boldsymbol{c} = 0 = |\boldsymbol{b}|.|\boldsymbol{c}|\cos\theta \Rightarrow \theta = 90°$.

(b) $\boldsymbol{d} \cdot \boldsymbol{e} = (3\hat{\boldsymbol{i}} - 2\hat{\boldsymbol{j}} - \hat{\boldsymbol{k}}) \cdot (\hat{\boldsymbol{i}} + \lambda\hat{\boldsymbol{j}} + 2\hat{\boldsymbol{k}}) = 3\hat{\boldsymbol{i}} \cdot \hat{\boldsymbol{i}} - 2\lambda\hat{\boldsymbol{j}} \cdot \hat{\boldsymbol{j}} - 2\hat{\boldsymbol{k}} \cdot \hat{\boldsymbol{k}} = 1 - 2\lambda$.

$$|\boldsymbol{d}| = \sqrt{3^2 + 2^2 + 1^2} = \sqrt{14}.$$
$$|\boldsymbol{e}| = \sqrt{1^2 + \lambda^2 + 2^2} = \sqrt{5 + \lambda^2}.$$
$$\therefore 1 - 2\lambda = \sqrt{14}\sqrt{5 + \lambda^2}\cos 90° = 0$$
$$\Rightarrow \quad 1 - 2\lambda = 0 \quad \Rightarrow \quad 1 = 2\lambda \quad \Rightarrow \quad \lambda = \tfrac{1}{2}.$$

5.6. $\boldsymbol{c} = \boldsymbol{a} + \boldsymbol{b} \Rightarrow \boldsymbol{c} \cdot \boldsymbol{c} = (\boldsymbol{a} + \boldsymbol{b}) \cdot (\boldsymbol{a} + \boldsymbol{b}) = \underbrace{\boldsymbol{a} \cdot \boldsymbol{a}}_{a^2} + \underbrace{2\boldsymbol{a} \cdot \boldsymbol{b}}_{2ab\cos\theta} + \underbrace{\boldsymbol{b} \cdot \boldsymbol{b}}_{b^2}$

We need to exercise some care here because the quantity $a \cdot b$ will yield an angle $\theta = 180° - C$ which in this example is an acute angle, rather than the obtuse angle required. Consequently, we must substitute $180° - C$ for θ, which therefore gives:

$$\boldsymbol{c} \cdot \boldsymbol{c} = a^2 + b^2 + 2ab\cos(180° - C) = a^2 + b^2 - 2ab\cos C.$$

5.7. (a) $\boldsymbol{r}_1 = a\hat{\boldsymbol{i}} - a\hat{\boldsymbol{j}} - a\hat{\boldsymbol{k}}$; $\boldsymbol{r}_2 = -a\hat{\boldsymbol{i}} + a\hat{\boldsymbol{j}} - a\hat{\boldsymbol{k}}$; $\boldsymbol{r}_3 = a\hat{\boldsymbol{i}} + a\hat{\boldsymbol{j}} + a\hat{\boldsymbol{k}}$; $\boldsymbol{r}_4 = -a\hat{\boldsymbol{i}} - a\hat{\boldsymbol{j}} + a\hat{\boldsymbol{k}}$.

(b) $|\boldsymbol{r}_3| = \sqrt{3a^2} = \sqrt{3}a \quad \Rightarrow \quad R = \sqrt{3}a \quad \Rightarrow \quad a = \dfrac{R}{\sqrt{3}}$.

(c) $\boldsymbol{r}_3 - \boldsymbol{r}_2 = (a\hat{\boldsymbol{i}} + a\hat{\boldsymbol{j}} + a\hat{\boldsymbol{k}}) - (-a\hat{\boldsymbol{i}} + a\hat{\boldsymbol{j}} - a\hat{\boldsymbol{k}}) = 2a\hat{\boldsymbol{i}} + 2a\hat{\boldsymbol{k}}$;

$$|\boldsymbol{r}_3 - \boldsymbol{r}_2| = \sqrt{(2a)^2 + (2a)^2} = \sqrt{8a^2} = \sqrt{8}a = \frac{\sqrt{8}R}{\sqrt{3}} = \frac{2\sqrt{2}R}{\sqrt{3}}.$$

5.8. (a) $\boldsymbol{a} \times \boldsymbol{c} = (2\hat{\boldsymbol{i}} + 3\hat{\boldsymbol{k}}) \times (\hat{\boldsymbol{i}} - 2\hat{\boldsymbol{j}} + \hat{\boldsymbol{k}}) = (-4\hat{\boldsymbol{i}} \times \hat{\boldsymbol{j}}) + (2\hat{\boldsymbol{i}} \times \hat{\boldsymbol{k}}) + (3\hat{\boldsymbol{k}} \times \hat{\boldsymbol{i}}) - (6\hat{\boldsymbol{k}} \times \hat{\boldsymbol{j}}) = -4\hat{\boldsymbol{k}} - 2\hat{\boldsymbol{j}} + 3\hat{\boldsymbol{j}} + 6\hat{\boldsymbol{i}} = 6\hat{\boldsymbol{i}} + \hat{\boldsymbol{j}} - 4\hat{\boldsymbol{k}}$.

(b) $\boldsymbol{c} \times \boldsymbol{a} = (\hat{\boldsymbol{i}} - 2\hat{\boldsymbol{j}} + \hat{\boldsymbol{k}}) \times (2\hat{\boldsymbol{i}} + 3\hat{\boldsymbol{k}}) = (3\hat{\boldsymbol{i}} \times \hat{\boldsymbol{k}}) + (-4\hat{\boldsymbol{j}} \times \hat{\boldsymbol{i}}) - (6\hat{\boldsymbol{j}} \times \hat{\boldsymbol{k}}) + (2\hat{\boldsymbol{k}} \times \hat{\boldsymbol{i}}) = -3\hat{\boldsymbol{j}} + 4\hat{\boldsymbol{k}} - 6\hat{\boldsymbol{i}} + 2\hat{\boldsymbol{j}} = -6\hat{\boldsymbol{i}} - \hat{\boldsymbol{j}} + 4\hat{\boldsymbol{k}}$.

(c) $|\boldsymbol{c} \times \boldsymbol{a}| = \sqrt{36 + 1 + 16} = \sqrt{53}$.

(d) $(\hat{\boldsymbol{i}} \times \hat{\boldsymbol{j}}) \times \hat{\boldsymbol{j}} = \hat{\boldsymbol{k}} \times \hat{\boldsymbol{j}} = -\hat{\boldsymbol{i}}$.

(e) $\hat{\boldsymbol{i}} \times (\hat{\boldsymbol{j}} \times \hat{\boldsymbol{j}}) = \hat{\boldsymbol{i}} \times \boldsymbol{0} = \boldsymbol{0}$.

5.9. (a) $\boldsymbol{a}\cdot\boldsymbol{b} = (a_1\hat{\boldsymbol{i}} + a_2\hat{\boldsymbol{j}} + a_3\hat{\boldsymbol{k}})\cdot(b_1\hat{\boldsymbol{i}} + b_2\hat{\boldsymbol{j}} + b_3\hat{\boldsymbol{k}})$
$= a_1b_1\hat{\boldsymbol{i}}\cdot\hat{\boldsymbol{i}} + a_2\, b_2\hat{\boldsymbol{j}}\cdot\hat{\boldsymbol{j}} + a_3b_3\hat{\boldsymbol{k}}\cdot\hat{\boldsymbol{k}} = a_1b_1 + a_2b_2 + a_3b_3.$

(b) $\boldsymbol{b}\times\boldsymbol{c} = (b_1\hat{\boldsymbol{i}} + b_2\hat{\boldsymbol{j}} + b_3\hat{\boldsymbol{k}})\times(c_1\hat{\boldsymbol{i}} + c_2\hat{\boldsymbol{j}} + c_3\hat{\boldsymbol{k}})$
$= b_1c_2\hat{\boldsymbol{i}}\times\hat{\boldsymbol{j}} + b_1c_3\hat{\boldsymbol{i}}\times\hat{\boldsymbol{k}} + b_2c_1\hat{\boldsymbol{j}}\times\hat{\boldsymbol{i}} + b_2c_3\hat{\boldsymbol{j}}\times\hat{\boldsymbol{k}} + b_3c_1\hat{\boldsymbol{k}}\times\hat{\boldsymbol{i}}$
$+ b_3c_2\hat{\boldsymbol{k}}\times\hat{\boldsymbol{j}}$
$= b_1c_2\hat{\boldsymbol{k}} - b_1c_3\hat{\boldsymbol{j}} - b_2c_1\hat{\boldsymbol{k}} + b_2c_3\hat{\boldsymbol{i}} + b_3c_1\hat{\boldsymbol{j}} - b_3c_2\hat{\boldsymbol{i}}$
$= (b_2c_3 - b_3c_2)\hat{\boldsymbol{i}} - (b_1c_3 - b_3c_1)\hat{\boldsymbol{j}} + (b_1c_2 - b_2c_1)\hat{\boldsymbol{k}}.$

5.10. (a) $\boldsymbol{a}\times\boldsymbol{b} = \begin{vmatrix}\hat{\boldsymbol{i}} & \hat{\boldsymbol{j}} & \hat{\boldsymbol{k}}\\ 1 & 1 & 1\\ 1 & -1 & 1\end{vmatrix} = \hat{\boldsymbol{i}}\begin{vmatrix}1 & 1\\ -1 & 1\end{vmatrix} - \hat{\boldsymbol{j}}\begin{vmatrix}1 & 1\\ 1 & 1\end{vmatrix} + \hat{\boldsymbol{k}}\begin{vmatrix}1 & 1\\ 1 & -1\end{vmatrix} = 2\hat{\boldsymbol{i}} - 2\hat{\boldsymbol{k}}.$

(b) $|\boldsymbol{a}\times\boldsymbol{b}| = \left|2\hat{\boldsymbol{i}} - 2\hat{\boldsymbol{k}}\right| = \sqrt{2^2 + 2^2} = \sqrt{8}.$

$\therefore$ Unit vector $= \dfrac{2}{\sqrt{8}}\hat{\boldsymbol{i}} - \dfrac{2}{\sqrt{8}}\hat{\boldsymbol{k}} = \dfrac{1}{\sqrt{2}}\hat{\boldsymbol{i}} - \dfrac{1}{\sqrt{2}}\hat{\boldsymbol{k}}.$

5.11. For $\boldsymbol{a} = a_2\hat{\boldsymbol{j}} + a_3\hat{\boldsymbol{k}},\ b = b_1\hat{\boldsymbol{i}},\ c = c_2\hat{\boldsymbol{j}}$:

$$\boldsymbol{a}\cdot(\boldsymbol{b}\times\boldsymbol{c}) = \begin{vmatrix}0 & a_2 & a_3\\ b_1 & 0 & 0\\ 0 & c_2 & 0\end{vmatrix} = a_3\begin{vmatrix}b_1 & 0\\ 0 & c_2\end{vmatrix} = a_3b_1c_2.$$

5.12. (a) $\boldsymbol{b} = b_1\hat{\boldsymbol{i}}$ and $|\boldsymbol{b}| = 0.600$ nm; $\boldsymbol{c} = c_2\hat{\boldsymbol{j}}$ and $|c| = 0.866$ nm:

$\therefore b_1 = 0.600$ nm and $c_2 = 0.866$ nm.

(b) $\boldsymbol{a}\cdot\boldsymbol{c} = (a_2\hat{\boldsymbol{j}} + a_3\hat{\boldsymbol{k}})\cdot(c_2\hat{\boldsymbol{j}}) = a_2c_2 = 0.824\times0.866\times\cos\beta\ \text{nm}^2$

$\Rightarrow a_2\times0.866$ nm $= 0.824\times0.866\times\cos\beta\ \text{nm}^2$
$\Rightarrow a_2 = 0.824\ \text{nm}\times\cos 122.9^\circ = -0.448$ nm.

(c) $|\boldsymbol{a}| = \sqrt{a_2^2 + a_3^2} = \sqrt{0.448^2 + a_3^2} = 0.824$ nm

$\Rightarrow 0.679 = 0.448^2 + a_3^2$
$\therefore a_3^2 = 0.679 - 0.2003 = 0.479\ \text{nm}^2.$
$\Rightarrow\ a_3 = \pm0.692$ nm.

(d) Volume of the unit cell is given by:

$$\boldsymbol{a}\cdot(\boldsymbol{b}\times\boldsymbol{c}) = a_3b_1c_2 = 0.692\times0.600\times0.866 = 0.36\ \text{nm}^3.$$

Subject Index